SpringerBriefs in Computer Science

Series Editors

Stan Zdonik
Peng Ning
Shashi Shekhar
Jonathan Katz
Xindong Wu
Lakhmi C. Jain
David Padua
Xuemin Shen
Borko Furht
V. S. Subrahmanian
Martial Hebert
Katsushi Ikeuchi
Bruno Siciliano

For further volumes:
http://www.springer.com/series/10028

Lingjie Duan · Jianwei Huang
Biying Shou

Cognitive Virtual Network Operator Games

 Springer

Lingjie Duan
Engineering Systems and Design Pillar
Singapore University of Technology
  and Design
Dover
Singapore

Biying Shou
Department of Management Sciences
City University of Hong Kong
Kowloon
Hong Kong SAR

Jianwei Huang
Department of Information Engineering
The Chinese University of Hong Kong
Shatin
Hong Kong SAR

ISSN 2191-5768   ISSN 2191-5776 (electronic)
ISBN 978-1-4614-8889-7   ISBN 978-1-4614-8890-3 (eBook)
DOI 10.1007/978-1-4614-8890-3
Springer New York Heidelberg Dordrecht London

Library of Congress Control Number: 2013949243

© The Author(s) 2013
This work is subject to copyright. All rights are reserved by the Publisher, whether the whole or part of the material is concerned, specifically the rights of translation, reprinting, reuse of illustrations, recitation, broadcasting, reproduction on microfilms or in any other physical way, and transmission or information storage and retrieval, electronic adaptation, computer software, or by similar or dissimilar methodology now known or hereafter developed. Exempted from this legal reservation are brief excerpts in connection with reviews or scholarly analysis or material supplied specifically for the purpose of being entered and executed on a computer system, for exclusive use by the purchaser of the work. Duplication of this publication or parts thereof is permitted only under the provisions of the Copyright Law of the Publisher's location, in its current version, and permission for use must always be obtained from Springer. Permissions for use may be obtained through Rights Link at the Copyright Clearance Center. Violations are liable to prosecution under the respective Copyright Law. The use of general descriptive names, registered names, trademarks, service marks, etc. in this publication does not imply, even in the absence of a specific statement, that such names are exempt from the relevant protective laws and regulations and therefore free for general use.
While the advice and information in this book are believed to be true and accurate at the date of publication, neither the authors nor the editors nor the publisher can accept any legal responsibility for any errors or omissions that may be made. The publisher makes no warranty, express or implied, with respect to the material contained herein.

Printed on acid-free paper

Springer is part of Springer Science+Business Media (www.springer.com)

Preface

Mobile virtual network operator (MVNO) is a secondary operator that obtains spectrum resources from the spectrum owner and provides its own wireless services to a niche market. As of today, there are more than 600 MVNOs worldwide and some of them are worth multi-billion US dollars. Many of them, however, face a severe spectrum supply problem: under the current static spectrum licensing policy, an MVNO cannot flexibly obtain spectrum resources to match the dynamic changes of market demands. With the help of the cognitive radio technology, a cognitive MVNO (or C-MVNO) is no longer stuck in a long-term leasing contract with the spectrum owner, and can acquire spectrum dynamically in short-term through both dynamic spectrum leasing and spectrum sensing.

This book provides an overview of C-MVNOs' decisions under investment flexibility, supply uncertainty, and market competition in cognitive radio networks. This is a new research area at the nexus of cognitive radio engineering and microeconomics. Our focus is an operator's joint spectrum investment and service pricing decisions. Compared to dynamic spectrum leasing, spectrum sensing is cheaper but would introduce supply uncertainty due to primary licensed users' stochastic traffic. The readers will learn how to tradeoff the two flexible investment choices under supply uncertainty. Furthermore, if there is more than one operator, we present analysis of the competition among operators in obtaining spectrum and pricing services to attract users.

The outline of this book is as follows. Chapter 1 explains why spectrum bands are scarce but under-utilized, and how the cognitive radio technology and its application (dynamic spectrum access) can resolve this paradox. Chapter 2 studies the optimal investment and pricing decisions of a single C-MVNO under spectrum supply uncertainty. Chapter 3 studies competitive C-MVNOs' optimal investment and pricing decisions, taking into account their heterogeneity in leasing costs and users' heterogeneity in wireless characteristics. Chapter 4 summarizes the main results in this book.

We would like to thank the series editor, Prof. Xuemin (Sherman) Shen from University of Waterloo, for encouraging us to prepare this monograph. We also want to thank Prof. Martin B. H. Weiss from University of Pittsburgh for his helpful comments from a very early stage of this line of work. Last but not least, we want to thank members of the Network Communications and Economics Lab

(NCEL) at the Chinese University of Hong Kong, for their constructive feedback during the past several years.

The work described in this book was supported by grants from the Research Grants Council of the Hong Kong Special Administrative Region, China (Project No. CUHK 412710, CUHK 412511, and CityU 144209), and the research grants from City University of Hong Kong (Project No. 7002517 and 7008116). It is also partially supported by the SUTD-MIT International Design Center (IDC) Grant at Singapore University of Technology and Design (Project No. IDSF1200106OH). Part of the results has appeared in our prior publications [1, 2, 3, 4] and in the first author's Ph.D. dissertation [5].

Singapore, 2013   Lingjie Duan
Hong Kong   Jianwei Huang
                                                                     Biying Shou

References

1. L. Duan, J. Huang, B. Shou, Competition with dynamic spectrum leasing, in *Proceedings of IEEE Symposium on New Frontiers in Dynamic Spectrum Access Networks (DySPAN)*, (2010)
2. L. Duan, J. Huang, B. Shou, Cognitive mobile virtual network operator: Investment andpricing with supply uncertainty, in *Proceedings of The IEEE International Conference on Computer Communications (INFOCOM)*, (2010)
3. L. Duan, J. Huang, B. Shou, Duopoly competition in dynamic spectrum leasing and pricing. IEEE Trans. Mob. Comput. **11**(11), 1706–1719, (2012). http://arxiv.org/abs/1003.5517
4. L. Duan, J. Huang, B. Shou, Investment and pricing with spectrum uncertainty: A cognitive operators perspective. IEEE Trans. Mob. Comput. **10**(11), (2011). http://arxiv.org/abs/0912.3089
5. L. Duan, *Some Economics of Cellular and Cognitive Radio Networks* (The Chinese University of Hong Kong, Hong Kong, 2012). Ph.D. Dissertation

Contents

Chapter 1
Overview

Abstract In this chapter, we will first explain the issues of the current static spectrum licensing approach of spectrum management. This is the source of the spectrum usage paradox: spectrum resource is scarce but much of it is severely under-utilized. This motivates the study of cognitive radio technology and dynamic spectrum access. Dynamic spectrum access aims to improve spectrum utilization through innovations in technology, economics, and policies. We will introduce different types of dynamic spectrum access, in particular, the dynamic exclusive use model and the hierarchical access model. Finally, we will provide a brief overview of the related work.

1.1 Issues of Static Spectrum Licensing

Radio frequency spectrum is the critical resource to support wireless transmissions, and only limited amount of spectrum can be efficiently used in wireless networks (e.g., 300 MHz–3 GHz for cellular networks). In today's wireless networks, frequency spectrum is regulated by governmental agencies (e.g., FCC in U.S. and Ofcom in U.K.) under the "command-and-control" spectrum management policy [1]. In such a static spectrum licensing approach, spectrum is allocated to license holders (e.g., primary licensed network operators) over large geographical areas for years or even decades. A primary network operator will use the licensed spectrum to exclusively serve his own primary licensed users. As a result, secondary unlicensed users cannot access the licensed bands due to lack of the proper licenses. Because of this, many people believe that we are running out of usable spectrum. This belief is further strengthened by operators' expensive bids for extra usable spectrum bands [15].[1] For example, in an European 3G spectrum auction for a merely 20 MHz spectrum band, the bid reached multi-billion dollars; in U.S. there were nearly 20 billion dollars netted for 700 MHz auction in 2008.

---

[1] To become a license holder of certain spectrum band, one needs to wait till the next round of spectrum auction organized by the governmental agency.

L. Duan et al., *Cognitive Virtual Network Operator Games*,
SpringerBriefs in Computer Science, DOI: 10.1007/978-1-4614-8890-3_1,
© The Author(s) 2013

Consider the increasing need of more spectrum to match users' ever-increasing data demands, one might wonder whether we are really close to the limit of spectrum resource. However, field spectrum occupancy measurements suggested otherwise [1, 2]. According to [1, 2], temporal and geographical variations in the utilization of the licensed spectrum rang from 15 to 85 %, and a large portion of licensed spectrum is severely under-utilized. A more recent measurement [3] shows that the overall average spectrum utilization does not exceed 20 % in densely populated cities such as Chicago and New York City. This paradox shows that the scarcity of spectrum resource is a result of the existing static spectrum licensing. To resolve this paradox, we need a more market-oriented and dynamic spectrum allocation approach.

1.2 Cognitive Radios and Dynamic Spectrum Access

1.2.1 Spectrum Opportunity and Cognitive Radios

Let us first understand the meaning of spectrum opportunity, which helps us to understand the key idea of cognitive radio later on. As a primary licensed user has the priority access right to the licensed spectrum, a spectrum band becomes a potential opportunity for secondary users if it is not used by primary users (if such access is allowed by the primary users) [15, 16]. For example, consider a secondary user (represented by a secondary transmitter-receiver pair) coexisting with geographically distributed primary users (primary transmitter-receiver pairs). The secondary user can safely utilize a channel in the spectrum if his transmitter will not introduce intolerant interference to primary receivers and his receiver does not suffer intolerant interferences from primary transmitters. This spectrum opportunity is thus a local definition, depending not only on the transmitter-receiver locations of the secondary and primary users, but also on the primary users' traffic activities.

Cognitive radio is the key technology for a secondary user to take advantage of the spectrum opportunities in a dynamic approach. Cognitive radio is context-aware intelligent radio that can change its transmission parameters according to the communication environment in which it operates [17]. Though secondary users have no spectrum licenses, they can share the licensed spectrum bands via additional functionalities enabled by cognitive radios. To catch a spectrum opportunity, cognitive radio enables a secondary party (i.e., secondary user or secondary operator representing the secondary users) to sense the varying radio environment and identify the unused portion of spectrum at a specific time and location. It should be noted that such a cognitive capacity is not limited to monitoring primary signals in spectrum bands, but also involves detecting temporal and spatial variations in radio environment and avoiding interference to primary receivers [4]. Specifically, a secondary party needs to carry out the following three tasks to dynamically detect and utilize the unused spectrum (i.e., spectrum holes or white space) [4, 5]:

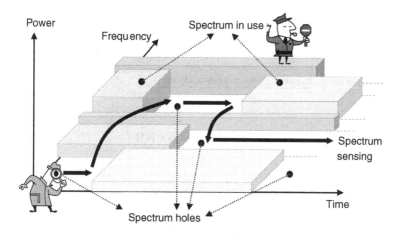

Fig. 1.1 Spectrum sensing for temporarily unused portion of spectrum (revised based on [4])

- *Spectrum sensing:* the secondary party monitors the radio environment over three domains at a particular location: primary signal power, frequency spectrum, and time horizon as in Fig. 1.1.[2] In this process, primary users are oblivious to the presence of the secondary party, and the secondary party needs to estimate the radio environment to avoid violating the usage rights of primary users [5, 24]
- *Spectrum analysis:* the secondary party then estimates the characteristics of the sensed spectrum holes (e.g., channel-state information and channel capacity for secondary use).
- *Spectrum decision:* the secondary party finally chooses the appropriate spectrum band and start transmission by specifying the transmission mode, transmission power, and data rate on the spectrum holes.

1.2.2 Three Typical Models in Dynamic Spectrum Access

Cognitive radio technology may fundamentally change many aspects of wireless communication networks, and is the key to enable dynamic spectrum access. Different from the traditional static spectrum licensing approach, dynamic spectrum access aims to achieve more efficient spectrum utilization by allowing the secondary users to opportunistically share the resource with the primary users. The success of dynamic spectrum access requires innovations in wireless engineering, economics, and policy. As shown in Fig. 1.2, we can roughly categorize the dynamic spectrum access into three models.

---

[2] There are multiple types of spectrum sensing techniques, e.g., the traditional energy detection which can be easily deployed, the cooperative detection to resolve hidden terminal problem, and the interference-based detection to better manage interference at the primary receivers [4].

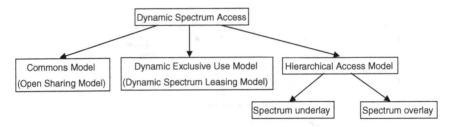

Fig. 1.2 Three categories of dynamic spectrum access as in [15]

1.2.2.1 Commons Model

The commons model is also referred to as the Open Sharing model. This model does not differentiate primary users and secondary users, and allows open sharing of spectrum among all users. One successful example is the industrial, scientific, and medical (ISM) spectrum band, which supports the very successful WiFi technology. However, without proper regulations, we will observe the "tragedy of commons" in such a model. This is because a selfish user will simply transmit at its maximum power to maximize his data rate in such a model. When everyone does this, the interferences become severe and no users can efficiently communicate. To respond to this challenge, researchers have proposed both centralized [7, 8] and distributed sharing schemes [12–14] to efficiently manage the common spectrum resource.

1.2.2.2 Dynamic Exclusive Use Model

Recall that in the current static spectrum licensing approach, spectrum bands are assigned to spectrum holders for exclusive use. On the contrary, dynamic exclusive use model (or dynamic spectrum leasing) introduces flexibility in improving the spectrum utilization. This approach allows the license holder to trade his spectrum band to another (usually unlicensed) party [20, 22, 24, 34, 90, 93]. For example, a primary network operator allows secondary users to operate in his temporarily unused part of spectrum in exchange of economic return (e.g., [22, 24]) or relay helps for inefficient primary transmissions (e.g,. [34, 110, 111]). The dynamic spectrum leasing can be short-term or even real-time (e.g., [35–37]), and can be at a similar time scale of the spectrum sensing operation. Note that the spectrum trading is not mandatory, and will only happen when the primary and secondary parties reach a win-win situation. As the exclusive holder of licensed spectrum, the primary party (operator or user) has more market power than the secondary party, and can often decide which collaboration scheme to propose. The main challenge for a primary party to properly incentivize such collaboration is that he may not know about secondary users' private information (e.g., valuations of exchanged bandwidth, local channel conditions, and local energy cost) and the collaboration scheme may need to be determined under asymmetric information [110, 111].

1.2.2.3 Hierarchical Access Model

The hierarchical access model allows a secondary party to opportunistically access the spectrum without affecting the normal operation of the primary users. There are two main approaches about primary and secondary users' hierarchical access structure: spectrum underlay [18, 19] and spectrum overlay [17, 42, 91]. The spectrum underlay approach allow concurrent transmission of the secondary and primary users in the same channel, but impose strict requirements on secondary users' transmission power so that they will not introduce intolerant interference to the primary users. One good example is the ultra-wide-band (UWB) system, where secondary users spread their transmission signals over a wide band and obtain high data rate in a short-range with a very low transmission power density. As this approach assumes a worst case where primary users always transmit, it does not need sophisticated detection scheme for spectrum holes.

Different from spectrum underlay, the spectrum overlay approach does not impose strict requirements on secondary users' transmission power but requires the avoidance of any possible collision with primary users' activities. It allows secondary users to exploit and detect temporal and spatial spectrum holes and opportunistically access in a non-intrusive way.

1.3 Related Research

In the following two chapters, we will focus on both the dynamic spectrum leasing model and the spectrum overlay approach of the hierarchical access model. We consider a secondary network operator that is equipped with cognitive radios (becoming cognitive network operator). It can flexibly obtain short-term spectrum resource by dynamically leasing from the spectrum owner and/or sensing the spectrum holes for usage. Both types of investment can be done at a short time scale, and the operator can jointly optimize his investment and pricing schemes to match market changes over time.

There is a growing interest in studying the investment and pricing decisions of cognitive network operators recently. Several auction mechanisms have been proposed to study the investment problems of cognitive network operators (e.g., [41, 43, 44]). Other recent results studied the pricing decisions of the cognitive network operators who interact with a group of secondary users (e.g., [45–53]). [41] considered users' queueing delays and obtained most results through simulations. [46] presented a recent survey on the spectrum sharing games of network operators and cognitive radio networks. [47] studied the competition among multiple service providers without modeling users' wireless details. [48] considered a pricing competition game of two operators and adopted a simplified wireless model for the users. [49] derived users' demand functions based on the acceptance probability model for the users. [50] explored demand functions based on both quality-sensitive and price-sensitive buyer population models. [51] formulated the interaction between one primary user (monopolist) and multiple secondary users as a Stackelberg game. The primary user uses some secondary users as relays and leases its bandwidth to those relays to collect revenue. [52] studied a multiple-level spectrum market among primary, secondary, and tertiary services where global information is not available. [53] considered the short-term spectrum trading between multiple primary users and multiple secondary users. The spectrum buying behaviors of secondary users are modeled as an evolutionary game, while selling behaviors of primary users are modeled as a noncooperative game. [49–53] obtained most interesting results through simulations. There are only few papers (e.g., [37, 52–54]) that jointly considered the spectrum investment and service pricing problem as Chaps. 2 and 3 (which are also reported in our papers [71, 80, 90, 93]). Moreover, none of these work considered spectrum sensing as a choice of investment, where the useful spectrum amount obtained through spectrum sensing is random due to primary licensed users' stochastic traffic. A key contribution of this book (in Chap. 2 and our papers [80, 93]) is to study impact of supply uncertainty due to spectrum sensing.

Our model of spectrum uncertainty in Chap. 2 is related to the random-yield model in supply chain management (e.g., [55–57]). In these work, a supplier (similar to the spectrum owner) provides a random output dependent of the order size of the retailer (similar to the cognitive operator). The unique wireless aspects of our system, however, make the analysis and insights very different.

Chapter 2
Secondary Spectrum Market Under Supply Uncertainty

Abstract This chapter studies the optimal investment and pricing decisions of a cognitive mobile virtual network operator (C-MVNO) under spectrum supply uncertainty. Compared with a traditional MVNO who is often stuck in long-term spectrum leasing contract, a C-MVNO can acquire spectrum dynamically in short-term by both sensing the empty "spectrum holes" of licensed bands and dynamically leasing from the spectrum owner. Compared to dynamic spectrum leasing, spectrum sensing is typically cheaper, but the obtained useful spectrum amount is random due to primary licensed users' stochastic traffic. The C-MVNO needs to determine the optimal amounts of spectrum sensing and leasing by evaluating the trade-off between cost and uncertainty. The C-MVNO also needs to determine the optimal price to sell the spectrum to the secondary unlicensed users, taking into account wireless heterogeneity of users such as different maximum transmission power levels and channel gains. We model and analyze the interactions between the C-MVNO and secondary unlicensed users as a Stackelberg game, and show interesting properties of the network equilibrium, including threshold structures of the optimal investment and pricing decisions.

2.1 Background

This chapter considers a secondary operator who obtains spectrum resource via both *spectrum sensing* for the spectrum overlay of hierarchical-access approach and *dynamic spectrum leasing* approach. In this chapter, we study the operation of a cognitive radio network that consists of a cognitive mobile virtual network operator (C-MVNO) and a group of secondary unlicensed users. The word "virtual" refers to the fact that the operator does not own the wireless spectrum bands or even the physical network infrastructure [72]. The C-MVNO serves as the interface between

L. Duan et al., *Cognitive Virtual Network Operator Games*,
SpringerBriefs in Computer Science, DOI: 10.1007/978-1-4614-8890-3_2,
© The Author(s) 2013

the primary operator and the secondary users which is similar to MVNO.[1] The word "cognitive" refers to the fact that the operator can obtain spectrum resource through both spectrum sensing using the cognitive radio technology [15, 24] and dynamic spectrum leasing from the primary operator [22, 24, 34]. The operator then resells the obtained spectrum (bandwidth) to secondary users to maximize its profit. The proposed model is a hybrid of the hierarchical-access and dynamic exclusive use models. It is applicable in various network scenarios, such as achieving efficient utilization of the TV spectrum in IEEE 802.22 standard [40]. This standard suggests that the secondary system should operate on a point-to-multipoint basis, i.e., the communications will happen between secondary base stations and secondary customer-premises equipment. The base stations can be operated by one or several C-MVNOs introduced in this chapter.

Compared with a traditional MVNO who only leases spectrum through long-term contracts, a C-MVNO can dynamically adjust its sensing and leasing decisions to match the changes of users' demand at a short time scale. Moreover, sensing often offers a cheaper way to obtain spectrum compared with leasing. The cost of sensing mainly includes the sensing time and energy, and does not include explicit cost paid to the primary operator. With a mature spectrum sensing technology, sensing cost should be reasonable low (otherwise there is no point of using cognitive radio). Spectrum leasing, however, involves direct negotiation with the primary operator. When the primary operator determines the cost of leasing, it needs to calculate its opportunity cost, i.e., how much revenue the spectrum can provide if the primary operator provides services directly over it. It is reasonable to believe that the leasing cost is more expensive than the sensing cost in most cases.[2] Although sensing is cheaper, the amount of spectrum obtained through sensing is often uncertain due to the stochastic nature of primary users' traffic. It is thus critical for a C-MVNO to find the right balance between cost and uncertainty.

Our key results and contributions are summarized as follows. For simplicity, we refer to the C-MVNO as "operator", secondary users as "users", and "dynamic leasing" as "leasing".

- A *Stackelberg game model*: We model and analyze the interactions between the operator and the users in the spectrum market as a Stackelberg game. As the leader, the operator makes the sensing, leasing, and pricing decisions sequentially. As the followers, users then purchase bandwidth from the operator to maximize their payoffs. By using backward induction, we prove the existence and uniqueness of the equilibrium, and show how various system parameters (i.e., sensing and leasing

---

[1] References [38, 39] show that it can be more efficient for the primary operator to hire an MVNO as intermediary to retail its spectrum resource, as MVNO can have a better understanding of local user population and users' demand. MVNOs can partially share the network investment cost and introduce new services as supplement to existing services provided by the primary operators [9]. Some regulators also wants primary operators to open their networks or resources to MVNOs such that more competition is introduced into the market [10].

[2] The analysis of this chapter also covers the case where sensing is more expensive than leasing, which is a trivial case to study.

costs, users' transmission power and channel conditions) affect the equilibrium behavior. Despite the complexity of the model, we are able to fully characterize the unique equilibrium behaviors of the operator and users.

- *Threshold structures of the optimal investment and pricing decisions*: At the equilibrium, the operator will sense the spectrum only if the sensing cost is cheaper than a threshold. Furthermore, it will lease some spectrum only if the resource obtained through sensing is below a threshold. Finally, the operator will charge a constant price to the users if the total bandwidth obtained through sensing and leasing does not exceed a threshold. The thresholds are easy to compute and the corresponding decisions rules are easy to implement in practice.
- *Fair and predictable QoS*: The operator's optimal pricing decision is independent of the users' wireless characteristics. Each user receives a payoff that is proportional to its channel gain and transmission power, which leads to the same signal-to-noise (SNR) for all users.
- *Impact of spectrum sensing*: We show that the availability of sensing always increases the operator's profit in the *expected* sense. The actual realization of the profit at a particular time heavily depends on the spectrum sensing results. Users always get better payoffs when the operator performs spectrum sensing.

Section 2.2 introduces the network model and problem formulation. In Sect. 2.3, we analyze the game model through backward induction. We discuss various insights obtained from the equilibrium analysis and present some numerical results in Sect. 2.4. In Sect. 2.5, we show the impact of spectrum sensing on both the operator and users. In Sect. 2.6, we extend our work to the incomplete information case, where the operator does not know about the distribution of sensing realization factor and needs to learn over time. We conclude in Sect. 2.7 and outline some future research directions.

2.2 Network Model

2.2.1 Background on Spectrum Sensing and Leasing

To illustrate the opportunity and trade-off of spectrum sensing and leasing, we consider a primary operator who divides its licensed spectrum into two types:[3]

- *Service Band*: This band is reserved for serving the spectrum owner's primary users (PUs). Since the PUs' traffic is stochastic, there will be some unused spectrum which changes dynamically. The operator can sense and utilize the unused portions. There are no explicit communications between the primary operator and the operator.

---

[3] Our model of dynamic spectrum leasing in transference band falls into the "exclusive-use" model, and spectrum sensing with opportunistic access falls into the "shared-use" model in [23]. Our model is a general combination of these well known models in literature.

- *Transference Band*: The primary operator temporarily does not use this band. The operator can lease the bandwidth through explicit communications with the primary operator. Since the transference band is not used for serving primary users, there are no "spectrum holes" and there is no need for sensing in this band.

Due to the short-term property of both sensing and leasing, the operator needs to make both the sensing and leasing decisions in each time slot.

The example in Fig. 2.1 demonstrates the dynamic opportunities for spectrum sensing, the uncertainty of sensing outcome, and the impact of sensing or leasing decisions. The primary operator's entire band is divided into small 34 channels.[4]

- Time slot 1: PUs use channels 1–4 and 11–15. The operator is unaware of this and senses channels 3–8. As a result, it obtains four unused channels (5–8). It leases additional 9 channels (20–28) from the transference band.
- Time slot 2: PUs change their behavior and use channels 1–6. The operator senses channels 5–14 and obtains eight unused channels (7–14). It leases additional five channels (23–27) from the transference band.

The choice of time slot length depends on characteristics of the primary traffic. The optimization of time slot length has been extensively studied in [58, 59, 105], where secondary users maximize their overall access time under the constraint that primary users should be sufficiently protected (e.g., the primary users' outage probability is below some threshold). In our simulations, we choose the length of time slot such that the probability that primary users' activities change within a time slot is very small. This ensures that the outage probability due to secondary users' access is tolerable to primary users.

2.2.2 Notations and Assumptions

We consider a cognitive network with one operator and a set $\mathscr{I} = \{1, \ldots, I\}$ of users. The operator has the cognitive capability and can sense the unused spectrum.

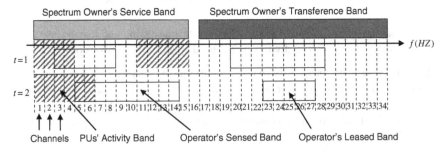

Fig. 2.1 Operator's investment in spectrum sensing and leasing

---

[4] Channel 16 is the guard band between the service and transference bands.

One way to realize this is to let the operator construct a sensor network that is dedicated to sensing the radio environment in space and time [60]. The operator will collect the sensing information from the sensor network and provide it to the unlicensed users, or providing "sensing as service". If the operator owns several base stations, then each base station is responsible for collecting sensing information in a certain geographical area. As mentioned in [60], there has been significant current research efforts in the context of an European project SENDORA [61], which aims at developing techniques based on sensor networks for supporting coexistence of licensed and unlicensed wireless users in a same area. The users are equipped with software defined radios and can tune to transmit in a wide range of frequencies as instructed by the operator, but do not necessarily have the cognitive sensing capacity.[5] Since the secondary users do not worry about sensing, they can spend most of their time and energy on actual data transmissions. Such a network structure puts most of the implementation complexity at the operator side and reduces the user equipment complexity, and thus might be easier to implement in practice than a "full" cognitive network.

The key notations of this chapter are listed in Table 2.1 with some explanations as follows.

- *Investment decisions B_s and B_l*: the operator's sensing and leasing bandwidths, respectively.

Table 2.1 Key notations

Symbol	Physical meaning
B_s	Sensing bandwidth
B_l	Leasing bandwidth
C_s	Unit sensing cost
C_l	Unit leasing cost
$\alpha \in [0, 1]$	Sensing realization factor
$\mathscr{I} = \{1, \cdots, I\}$	Set of secondary users
π	Unit price
w_i	User i's bandwidth allocation
r_i	User i's data rate
$P_i^{\max}$	User i's maximum transmission power
h_i	User i's channel gain
n_0	Noise power density
$g_i = P_i^{\max} h_i / n_0$	User i's wireless characteristic
$\text{SNR}_i = g_i / w_i$	User i's SNR
$G = \sum_{i \in \mathscr{I}} g_i$	Users' aggregate wireless characteristics
R	Operator's profit

[5] Even with the cognitive sensing capability, a secondary user may suffer from poor detection performance such a high missed detection probability. The sensor network infrastructure established by the operator can realize space diversity and reach good detection performance [30].

- *Sensing realization factor* α: when the operator senses a total bandwidth of B_s in a time slot, only a proportion of $\alpha \in [0, 1]$ is unused and can be used by the operator. α is a random variable and depends on the primary users' activities. With perfect sensing results, users can use bandwidth up to $B_s\alpha$ without generating interferences to the primary users. We first consider the case where the operator already knows the distribution of α.[6] We will relax this assumption in Sect. 2.6 and discuss how the operator learn α distribution over time.

- *Cost parameters* C_s *and* C_l: the operator's fixed sensing and leasing costs per unit bandwidth, respectively. Sensing cost C_s depends on the operator's sensing technologies. We focus on the commonly used energy detection for sensing technology [15]. To track and measure the energy of received signal, the operator needs to use a bandpass filter to square the output signal and then integrate over a proper observation interval. Thus the sensing cost only involves time and energy spent on channel sampling and signal processing [11, 58]. Sensing over different channels often needs to be done sequentially due to the potentially large number of channels open to opportunistic spectrum access and the limited power/hardware capacity of cognitive radios [66]. The larger sensing bandwidth and the more channels, the longer time and higher energy it requires [67]. For simplicity, we assume that total sensing cost is linear in the sensing bandwidth B_s. Leasing cost C_l is determined through the negotiation between the operator and the primary operator and is assumed to be larger than C_s.[7]

- *Pricing decision* π: the operator's choice of price per unit bandwidth to the users.

2.2.3 A Stackelberg Game

We consider a Stackelberg Game between the operator and the users as shown in Fig. 2.2. The operator is the Stackelberg leader: it first decides the sensing amount B_s in Stage I, then decides the leasing amount B_l in Stage II (based on the sensing result $B_s\alpha$), and then announces the price π to the users in Stage III (based on the total supply $B_s\alpha + B_l$). Finally, the users choose their bandwidth demands to maximize their individual payoffs in Stage IV.

We note that "sensing followed by leasing and pricing" is optimal for the operator to maximize its profit. Assuming sensing (though unreliable) is cheaper than leasing, the operator should observe sensing result first and then lease only if sensing does not provide enough resource. If the operator determines sensing, leasing and pricing simultaneously, then it is likely to "over-lease" expensive resource (compared with "sensing followed by leasing") to avoid having too little resource when α is small.

---

[6] This is reasonable if the operator can extensively measure PUs' activity patterns beforehand [62, 63], and then approximate the α distribution accurately as in [64, 65].

[7] If C_l is smaller than C_s, then the case becomes trivial as the operator will only lease spectrum. In a more general model, the primary operator can choose the value of C_l to maximize its own profit. We will study this model in our future work.

Fig. 2.2 A Stackelberg game

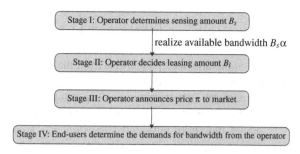

Also, simultaneously determination of price makes it harder to reach the market equilibrium where supply equals demand.

We can also show that optimizing the leasing and pricing decisions sequentially (as in our chapter) leads to the same profit if we optimize them simultaneously.

2.3 Backward Induction of the Four-Stage Game

The Stackelberg game falls into the class of dynamic game, and the common solution concept is the Subgame Perfect Equilibrium (SPE, or simply as *equilibrium* in this chapter). Note that the traditional Nash equilibrium investigates players' simultaneous actions in static game, thus is not applicable to our dynamic model [68]. A general technique for determining the SPE is the backward induction [69]. We will start with Stage IV and analyze the users' behaviors given the operator's investment and pricing decisions. Then we will look at Stage III and analyze how the operator makes the pricing decision given investment decisions and the possible reactions of the users in Stage IV. Finally we proceed to derive the operator's optimal leasing decision in Stage II and then the optimal sensing decision in Stage I. The backward induction captures the sequential dependence of the decisions in four stages.

2.3.1 Spectrum Allocation in Stage IV

In Stage IV, end-users determine their bandwidth demands given the unit price π announced by the operator in stage III. Each user can represent a transmitter–receiver node pair in an ad hoc network, or a node that transmits to the operator's base station in an uplink scenario. We assume that users access the spectrum provided by the operator through OFDM (Orthogonal frequency-division multiplexing) to avoid mutual interferences.[8] User i's achievable rate (in nats) is:[9]

---

[8] We focus on a single OFDMA cell case, where users transmit over orthogonal bands. The interference management across multiple cells is beyond the scope of this chapter.

[9] We assume that the operator only provides bandwidth without restricting the application types. This assumption has been commonly used in dynamic spectrum sharing literature, e.g., [34, 37, 47, 53].

$$r_i(w_i) = w_i \ln\left(1 + \frac{P_i^{\max} h_i}{n_0 w_i}\right),\tag{2.1}$$

where w_i is the allocated bandwidth from the operator, $P_i^{\max}$ is user i's maximum transmission power, n_0 is the noise power per unit bandwidth, h_i is user i's channel gain between user i's transmitter to the operator's secondary base station in an uplink scenario. To obtain rate in (2.1), user i spreads its maximum transmission power $P_k^{\max}$ across the entire allocated bandwidth w_i. To simplify the notation, we let $g_i = P_i^{\max} h_i / n_0$, thus g_i / w_i is the user i's signal-to-noise ratio (SNR). Here we focus on best-effort users who are interested in maximizing their data rates. Each user only knows its local information (i.e., $P_i^{\max}$, h_i, and n_0) and does not know anything about other users.

From a user's point of view, it does not matter whether the bandwidth has been obtained by the operator through spectrum sensing or dynamic leasing. Each unit of allocated bandwidth is perfectly reliable for the user.

To obtain closed-form solutions, we *first focus on the high SNR regime where* SNR $\gg 1$. This is motivated by the fact that users often have limited choices of modulation and coding schemes, and thus may not be able to decode a transmission if the SNR is below a threshold. In the high SNR regime, the rate in (2.1) can be approximated as

$$r_i(w_i) = w_i \ln\left(\frac{g_i}{w_i}\right).\tag{2.2}$$

Although the analytical solutions in Sect. 2.3 are derived based on (2.2), we emphasize that *all the major engineering insights remain true in the general SNR regime.* A formal proof is in Sect 2.4.

A user i's *payoff* is a function of the allocated bandwidth w_i and the price π,

$$u_i(\pi, w_i) = w_i \ln\left(\frac{g_i}{w_i}\right) - \pi w_i,\tag{2.3}$$

i.e., the difference between the data rate and the linear payment (πw_i). Payoff $u_i(\pi, w_i)$ is concave in w_i, and the unique bandwidth *demand* that maximizes the payoff is

$$w_i^*(\pi) = \arg\max_{w_i \geq 0} u_i(\pi, w_i) = g_i e^{-(1+\pi)},\tag{2.4}$$

which is always positive, linear in g_i, and decreasing in price π. Since g_i is linear in channel gain h_i and transmission power $P_i^{\max}$, then a user with a better channel condition or a larger transmission power has a larger demand.

Equation (2.4) shows that each user i achieves the same SNR:

$$\text{SNR}_i = \frac{g_i}{w_i^*(\pi)} = e^{1+\pi}.$$

but a different payoff that is linear in g_i,

$$u_i(\pi, w_i^*(\pi)) = g_i e^{-(1+\pi)}.$$

We denote users' aggregate wireless characteristics as $G = \sum_{i \in \mathscr{I}} g_i$. The users' total demand is

$$\sum_{i \in \mathscr{I}} w_i^*(\pi) = G e^{-(1+\pi)}. \qquad (2.5)$$

Next, we consider how the operator makes the investment (sensing and leasing) and pricing decisions in Stages I-III based on the total demand in Eq. (2.5).[10] In particular, we will show that the operator will always choose a price in Stage III such that the total demand (as a function of price) does not exceed the total supply.

2.3.2 Optimal Pricing Strategy in Stage III

We focus on the uplink transmissions in an infrastructure based secondary network (like the one proposed in IEEE 802.22 standard), where the secondary users need to communicate directly with the secondary base station (i.e., the operator). Similar as today's cellular network, a user needs to register with the operator when it enters and leaves the network. Thus at any given time, the operator knows precisely how many users are using the service. Equation (2.4) shows that each user's demand depends on the received power (i.e., the product of its transmission power and the channel gain) at the secondary base station in the uplink cellular network. This can be measured at the secondary base station when the user first registers with the operator. Thus the operator knows the exact demand from the users as well as user population in our model.

In Stage III, the operator determines the optimal pricing considering users' total demand (2.5), given the bandwidth supply $B_s \alpha + B_l$ obtained in Stage II. The operator profit is

$$R(B_s, \alpha, B_l, \pi) = \min\left(\pi \sum_{i \in \mathscr{I}} w_i^*(\pi), \pi \left(B_l + B_s \alpha\right)\right) - (B_s C_s + B_l C_l), \quad (2.6)$$

which is the difference between the revenue and total cost. The min operation denotes the fact that the operator can only satisfy the demand up to its available supply. The objective of Stage III is to find the optimal price $\pi^*(B_s, \alpha, B_l)$ that maximizes the profit, that is,

$$R_{III}(B_s, \alpha, B_l) = \max_{\pi \geq 0} R(B_s, \alpha, B_l, \pi). \qquad (2.7)$$

The subscript "III" denotes the best profit in Stage III.

_____

[10] We assume that the operator knows the value of G through proper feedback mechanism from the users.

Since the bandwidths B_s and B_l are given in this stage, the total cost $B_s C_s + B_l C_l$ is already fixed. The only optimization is to choose the optimal price π to maximize the revenue, i.e.,

$$\max_{\pi \geq 0} \min \left(\pi \sum_{i \in \mathscr{I}} w_i^*(\pi), \pi \, (B_l + B_s \alpha) \right). \tag{2.8}$$

The solution of problem (2.8) depends on the bandwidth investment in Stages I and II. Let us define $D(\pi) = \pi \sum_{i \in \mathscr{I}} w_i^*(\pi)$ and $S(\pi) = \pi(B_l + B_s\alpha)$. Figure 2.3 shows three possible relationships between these two terms, where $S_j(\pi)$ (for $j = 1, 2, 3$) represents each of the three possible choices of $S(\pi)$ depending on the bandwidth $B_l + B_s\alpha$:

- $S_1(\pi)$ (excessive supply): No intersection with $D(\pi)$;
- $S_2(\pi)$ (excessive supply): intersect once with $D(\pi)$ where $D(\pi)$ has a non-negative slope;
- $S_3(\pi)$ (conservative supply): intersect once with $D(\pi)$ where $D(\pi)$ has a negative slope.

In the *excessive supply* regime, $\max_{\pi \geq 0} \min (S(\pi), D(\pi)) = \max_{\pi \geq 0} D(\pi)$, i.e., the max-min solution occurs at the maximum value of $D(\pi)$ with $\pi^* = 1$. In this regime, the total supply is larger than the total demand at the best price choice. In the *conservative supply* regime, the max-min solution occurs at the unique intersection point of $D(\pi)$ and $S(\pi)$. The above observations lead to the following result.

Theorem 2.1. *The optimal pricing decision and the corresponding optimal profit at Stage III can be characterized by Table 2.2.*

The proof of Theorem 2.1 is given in Appendix 2.8.1. Note that in the excessive supply regime, some bandwidth is left unsold (i.e., $S(\pi^*) > D(\pi^*)$). This is because

Fig. 2.3 Different intersection cases of $D(\pi)$ and $S(\pi)$

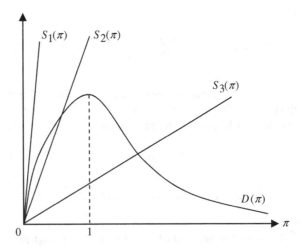

Table 2.2 Optimal pricing decision and profit in stage III

Total bandwidth obtained in stages I & II	Optimal price $\pi^*(B_s, \alpha, B_l)$	Optimal profit $R_{III}(B_s, \alpha, B_l)$
Excessive supply regime: $B_l + B_s\alpha \geq Ge^{-2}$	$\pi^{ES} = 1$	$R_{III}^{ES}(B_s, \alpha, B_l) = Ge^{-2} - B_sC_s - B_lC_l$
Conservative supply regime: $B_l + B_s\alpha < Ge^{-2}$	$\pi^{CS} = \ln\left(\frac{G}{B_l+B_s\alpha}\right) - 1$	$R_{III}^{CS}(B_s, \alpha, B_l) = (B_l + B_s\alpha)$ $\ln\left(\frac{G}{B_l+B_s\alpha}\right) - B_s(\alpha + C_s) - B_l(1 + C_l)$

the acquired bandwidth is too large, and selling all the bandwidth will lead to a very low price that decreases the revenue (the product of price and sold bandwidth). The profit can be apparently improved if the operator acquires less bandwidth in Stages I and II. Later analysis in Stages II and I will show that the equilibrium of the game must lie in the conservative supply regime if the sensing cost is non-negligible. In the conservative supply regime, the optimal price still guarantees supply equal to demand (i.e., market equilibrium).

2.3.3 Optimal Leasing Strategy in Stage II

In Stage II, the operator decides the optimal leasing amount B_l given the sensing result $B_s\alpha$:

$$R_{II}(B_s, \alpha) = \max_{B_l \geq 0} R_{III}(B_s, \alpha, B_l). \tag{2.9}$$

We decompose problem (2.9) into two subproblems based on the two supply regimes in Table 2.2,

1. Choose B_l to reach the excessive supply regime in Stage III:

$$R_{II}^{ES}(B_s, \alpha) = \max_{B_l \geq \max\{Ge^{-2}-B_s\alpha, 0\}} R_{III}^{ES}(B_s, \alpha, B_l). \tag{2.10}$$

2. Choose B_l to reach the conservative supply regime in Stage III:

$$R_{II}^{CS}(B_s, \alpha) = \max_{0 \leq B_l \leq Ge^{-2}-B_s\alpha} R_{III}^{CS}(B_s, \alpha, B_l), \tag{2.11}$$

To solve subproblems (2.10) and (2.11), we need to consider the bandwidth obtained from sensing.

- *Excessive Supply* $(B_s\alpha > Ge^{-2})$: in this case, the feasible sets of both subproblems (2.10) and (2.11) are empty. In fact, the bandwidth supply is already in the excessive supply regime as defined in Table II, and it is optimal not to lease in Stage II.

- *Conservative Supply* $(B_s \alpha \leq Ge^{-2})$: first, we can show that the unique optimal solution of subproblem (2.10) is $B_l^* = Ge^{-2} - B_s \alpha$. This means that the optimal objective value of subproblem (2.10) is no larger than that of subproblem (2.11), and thus it is enough to consider subproblem (2.11) in the conservative supply regime only.

Base on the above observations and some further analysis, we can show the following:

Theorem 2.2. *In Stage II, the optimal leasing decision and the corresponding optimal profit are summarized in Table 2.3.*

The proof of Theorem 2.2 is given in Appendix 2.8.2. Table 2.3 contains three cases based on the value of $B_s \alpha$: (CS1), (CS2), and (ES3). The first two cases involve solving the subproblem (2.11) in the conservative supply regime, and the last one corresponds to the excessive supply regime. Although the decisions in cases (CS2) and (ES3) are the same (i.e., zero leasing amount), we still treat them separately since the profit expressions are different.

It is clear that we have an optimal *threshold* leasing policy here: the operator wants to achieve a total bandwidth equal to $Ge^{-(2+C_l)}$ whenever possible. When the bandwidth obtained through sensing is not enough, the operator will lease additional bandwidth to reach the threshold; otherwise the operator will not lease.

2.3.4 Optimal Sensing Strategy in Stage I

In Stage I, the operator will decide the optimal sensing amount to maximize its expected profit by taking the uncertainty of the sensing realization factor α into account. The operator needs to solve the following problem

$$R_I = \max_{B_s \geq 0} R_{II}(B_s),$$

Table 2.3 Optimal leasing decision and profit in stage II

Given sensing result $B_s \alpha$ After Stage I	Optimal leasing amount B_l^*	Optimal profit $R_{II}(B_s, \alpha)$	
(CS1) $B_s \alpha \leq Ge^{-(2+C_l)}$	$B_l^{CS1} = Ge^{-(2+C_l)} - B_s \alpha$	$R_{II}^{CS1}(B_s, \alpha) =$ $Ge^{-(2+C_l)}$ $+$ $B_s(\alpha C_l - C_s)$	
(CS2) $B_s \alpha \in \left(Ge^{-(2+C_l)}, Ge^{-2}\right]$	$B_l^{CS2} = 0$	$R_{II}^{CS2}(B_s, \alpha) =$ $B_s \alpha \ln\left(\frac{G}{B_s \alpha}\right) -$ $B_s(\alpha + C_s)$	
(ES3) $B_s \alpha > Ge^{-2}$	$B_l^{ES3} = 0$	$R_{II}^{ES3}(B_s, \alpha) =$ $Ge^{-2} - B_s C_s$	

where $R_{II}(B_s)$ is obtained by taking the expectation of α over the profit functions in Stage II (i.e., $R_{II}^{CS1}(B_s, \alpha)$, $R_{II}^{CS2}(B_s, \alpha)$, and $R_{II}^{ES3}(B_s, \alpha)$ in Table 2.3).

To obtain closed-form solutions, we assume that the sensing realization factor α follows a uniform distribution in $[0, 1]$. In Sect. 2.4.1, we prove that *the major engineering insights also hold under any general distribution*.

To avoid the trivial case where sensing is so cheap that it is optimal to sense a huge amount of bandwidth, we further assume that the sensing cost is non-negligible and is lower bounded by $C_s \geq (1 - e^{-2C_l})/4$.

To derive function $R_{II}(B_s)$, we will consider the following three intervals:

1. Case I: $B_s \in [0, Ge^{-(2+C_l)}]$. In this case, we always have $B_s\alpha \leq Ge^{-(2+C_l)}$ for any value $\alpha \in [0, 1]$, which corresponds to case (CS1) in Table 2.3. The expected profit is

$$R_{II}^1(B_s) = E_{\alpha \in [0,1]}\left[R_{II}^{CS1}(B_s, \alpha)\right] = Ge^{-(2+C_l)} + B_s\left(\frac{C_l}{2} - C_s\right),$$

   which is a linear function of B_s. If $C_s > C_l/2$, $R_{II}^1(B_s)$ is linearly decreasing in B_s; if $C_s < C_l/2$, $R_{II}^1(B_s)$ is linearly increasing in B_s.

2. Case II: $B_s \in \left(Ge^{-(2+C_l)}, Ge^{-2}\right]$. Depending on the value of α, $B_s\alpha$ can be in either case (CS1) or case (CS2) in Table 2.3. The expected profit is

$$R_{II}^2(B_s) = E_{\alpha \in \left[0, \frac{Ge^{-(2+C_l)}}{B_s}\right]}\left[R_{II}^{CS1}(B_s, \alpha)\right] + E_{\alpha \in \left[\frac{Ge^{-(2+C_l)}}{B_s}, 1\right]}\left[R_{II}^{CS2}(B_s, \alpha)\right]$$

$$= \frac{B_s}{2}\ln\left(\frac{G}{B_s}\right) - \frac{B_s}{4} + \frac{B_s}{4}\left(\frac{Ge^{-(2+C_l)}}{B_s}\right)^2 - B_sC_s.$$

   $R_{II}^2(B_s)$ is a strictly concave function of B_s since its second-order derivative

$$\frac{\partial^2 R_{II}^2(B_s)}{\partial B_s^2} = \frac{1}{2B_s}\left[\left(\frac{Ge^{-(2+C_l)}}{B_s}\right)^2 - 1\right] < 0$$

   as $B_s > Ge^{-(2+C_l)}$ in this case.

3. Case III: $B_s \in \left(Ge^{-2}, \infty\right)$. Depending on the value of α, $B_s\alpha$ can be any of the three cases in Table 2.3. The expected profit is

$$R_{II}^3(B_s) = E_{\alpha \in \left[0, \frac{Ge^{-(2+C_l)}}{B_s}\right]}\left[R_{II}^{CS1}(B_s, \alpha)\right] + E_{\alpha \in \left[\frac{Ge^{-(2+C_l)}}{B_s}, \frac{Ge^{-2}}{B_s}\right]}\left[R_{II}^{CS2}(B_s, \alpha)\right]$$

$$+ E_{\alpha \in \left[\frac{Ge^{-2}}{B_s}, 1\right]}\left[R_{II}^{ES3}(B_s, \alpha)\right]$$

$$= \left(\frac{G}{e^2}\right)^2\frac{e^{-2C_l} - 1}{4B_s} - B_sC_s + \frac{G}{e^2}.$$

Because its first-order derivative

$$\frac{\partial R_{II}^3(B_s)}{\partial B_s} = \left(\frac{Ge^{-2}}{B_s}\right)^2 \frac{1 - e^{-2C_l}}{4} - C_s < 0,$$

as $B_s > Ge^{-2}$ in this case, $R_{II}^3(B_s)$ is decreasing in B_s and achieves its maximum at $B_s = Ge^{-2}$.

To summarize, the operator needs to maximize

$$R_{II}(B_s) = \begin{cases} R_{II}^1(B_s), & \text{if } 0 \le B_s \le Ge^{-(2+C_l)}; \\ R_{II}^2(B_s), & \text{if } Ge^{-(2+C_l)} < B_s \le Ge^{-2}; \\ R_{II}^3(B_s), & \text{if } B_s > Ge^{-2}. \end{cases} \tag{2.12}$$

We can verify that Case II always achieves a higher optimal profit than Case III. This means that the optimal sensing will only lead to either case (CS1) or case (CS2) in Stage II, which corresponds to the conservative supply regime in Stage III. This confirms our previous intuition that equilibrium is always in the conservative supply regime under a non-negligible sensing cost, since some resource is wasted in the excessive supply regime (see discussions in Sect. 2.3.2).

Table 2.4 shows that the sensing decision is made in the following two cost regimes:

- *High sensing cost regime* ($C_s > C_l/2$): it is optimal not to sense. Intuitively, the coefficient $1/2$ is due to the uniform distribution assumption of α, i.e., on average obtaining one unit of available bandwidth through sensing costs $2C_s$.
- *Low sensing cost regime* ($C_s \in \left[\frac{1-e^{-2C_l}}{4}, \frac{C_l}{2}\right)$): the optimal sensing amount B_s^{L*} is the unique solution to the following equation:

$$\frac{\partial R_{II}^2(B_s)}{\partial B_s} = \frac{1}{2} \ln\left(\frac{1}{B_s/G}\right) - \frac{3}{4} - C_s - \left(\frac{e^{-(2+C_l)}}{2B_s/G}\right)^2 = 0. \tag{2.13}$$

The uniqueness of the solution is due to the strict concavity of $R_{II}^2(B_s)$ over B_s. We can further show that B_s^{L*} lies in the interval of $\left[Ge^{-(2+C_l)}, Ge^{-2}\right]$ and is linear in G. Finally, the operator's optimal expected profit is

$$R_I^L = \frac{B_s^{L*}}{2} \ln\left(\frac{G}{B_s^{L*}}\right) - \frac{B_s^{L*}}{4} + \frac{1}{4B_s^{L*}}\left(\frac{G}{e^{2+C_l}}\right)^2 - B_s^{L*}C_s. \tag{2.14}$$

Based on these observations, we can show the following:

Theorem 2.3. *In Stage I, the optimal sensing decision and the corresponding optimal profit are summarized in Table 2.4. The optimal sensing amount B_I^* is linear in G.*

Table 2.4 Choice of optimal sensing amount in stage I

	Optimal sensing decision B_s^*	Expected Profit R_I
High sensing cost regime: $C_s \geq C_l/2$	$B_s^* = 0$	$R_I^H = Ge^{-(2+C_l)}$
Low Sensing Cost Regime: $C_s \in [(1 - e^{-2C_l})/4, C_l/2]$	$B_s^* = B_s^{L*}$, solution to Eq. (2.13)	R_I^L in Eq. (2.14)

Fig. 2.4 Expected profit in Stage II under different sensing and leasing costs

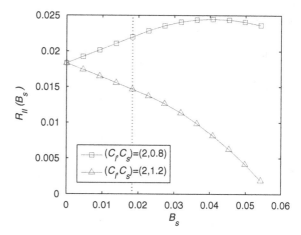

Figure 2.4 shows two possible cases for the function $R_{II}(B_s)$. The vertical dashed line represents $B_s = e^{-(2+C_l)}$. For illustration purpose, we assume $G = 1$, $C_l = 2$, and $C_s = \{0.8, 1.2\}$. When the sensing cost is large (i.e., $C_s = 1.2 > C_l/2$), $R_{II}(B_s)$ achieves its optimum at $B_s = 0$ and thus it is optimal not to sense. When the sensing cost is small (i.e., $C_s = 0.8 < C_l/2$), $R_{II}(B_s)$ achieves its optimum at $B_s > e^{-(2+C_l)}$ and it is optimal to sense a positive amount of spectrum.

2.4 Equilibrium Summary and Numerical Results

Based on the discussions in Sect. 2.3, we summarize the operator's equilibrium sensing/leasing/pricing decisions and the equilibrium resource allocations to the users in Table 2.5. These decisions can be directly computed by the operator in each time slot without using any iterative algorithm.

Several interesting observations are as follows.

Observation 1 *Both the optimal sensing amount B_s^* (either 0 or B_s^{L*}) and leasing amount B_l^* are linear in the users' aggregate wireless characteristics $G = \sum_{i \in \mathscr{I}} P_i^{max} h_i / n_0$.*

Table 2.5 The operator's and users' equilibrium behaviors

Sensing cost regimes	High sensing cost: $C_s \geq \frac{C_l}{2}$	Low sensing cost: $\frac{1-e^{-2C_l}}{4} \leq C_s \leq \frac{C_l}{2}$	
Optimal sensing amount B_s^*	0	$B_s^{L*} \in \left[Ge^{-(2+C_l)}, Ge^{-2} \right]$, solution to Eq. (2.14)	
Sensing realization factor α	$0 \leq \alpha \leq 1$	$0 \leq \alpha \leq Ge^{-(2+C_l)}/B_s^{L*}$	$\alpha > Ge^{-(2+C_l)}/B_s^{L*}$
Optimal leasing amount B_l^*	$Ge^{-(2+C_l)}$	$Ge^{-(2+C_l)} - B_s^{L*}\alpha$	0
Optimal price π^*	$1 + C_l$	$1 + C_l$	$\ln\left(\frac{G}{B_s^{L*}\alpha}\right) - 1$
Expected profit R_l	R_l^H $Ge^{-(2+C_l)}$	$=$ R_l^L in Eq. (2.14)	R_l^L in Eq. (2.14)
User i's SNR	$e^{(2+C_l)}$	$e^{(2+C_l)}$	$\frac{G}{B_s^{L*}\alpha}$
User i's Payoff	$g_i e^{-(2+C_l)}$	$g_i e^{-(2+C_l)}$	$g_i(B_s^{L*}\alpha/G)$

Fig. 2.5 Optimal sensing amount B_s^* as a function of C_s and C_l

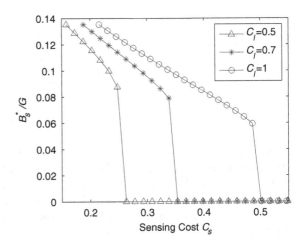

The linearity enables us to normalize optimal sensing and leasing decisions by users' aggregate wireless characteristics, and study the relationships between the normalized optimal decisions and other system parameters as in Figs. 2.5 and 2.6.

Figure 2.5 shows how the normalized optimal sensing decision B_s^*/G changes with the costs. For a given leasing cost C_l, the optimal sensing decision B_s^* decreases as the sensing cost C_s becomes more expensive, and drops to zero when $C_s \geq C_l/2$. For a given sensing cost C_s, the optimal sensing decision B_s^* increases as the leasing cost C_l becomes more expensive, in which case sensing becomes more attractive. Note that the sensing decision B_s^* is the same in each time slot if the users' population and channel conditions do not change.

Figure 2.6 shows how the normalized optimal leasing decision B_l^*/G depends on the costs C_l and C_s as well as the sensing realization factor α in the low sensing

Fig. 2.6 Optimal leasing amount B_l^* as a function of C_s, C_l, and α

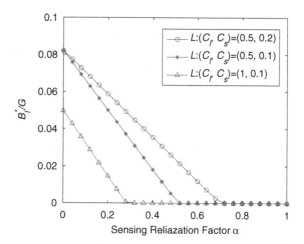

cost regime (denoted by "L"). In all cases, a higher value α means more bandwidth is obtained from sensing and there is a less need to lease. Fig. 2.6 confirms the threshold structure of the optimal leasing decisions in Sect. 2.3.3, i.e., no leasing is needed whenever the bandwidth obtained from sensing reaches a threshold. Comparing different curves, we can see that the operator chooses to lease more as leasing becomes cheaper or sensing becomes more expensive. For high sensing cost regime, the optimal leasing amount only depends on C_l and is independent of C_s and α, and thus is not shown here. Note that the leasing decision B_l^* may change with the sensing realization factor α, which depends on the burstiness of the primary user's stochastic traffic.

Observation 2 *The optimal pricing decision π^* in Stage III is independent of users' aggregate wireless characteristics G.*

Observation 2 is closely related to Observation 1. Since the total bandwidth is linear in G, the "average" resource allocation per user is "constant" at the equilibrium. This implies that the price must be independent of the user population change, otherwise the resource allocation to each individual user will change with the price accordingly.

Observation 3 *The optimal pricing decision π^* in Stage III is non-increasing in α in the low sensing cost regime.*

First, in the low sensing cost regime where the sensing result is poor (i.e., α is small as the third column in Table 2.5), the operator will lease additional resource such that the total bandwidth reaches the threshold $Ge^{-(2+C_l)}$. In this case, the price is a constant and is independent of the value of α. Second, when the sensing result is good (i.e., α is large as in the last column in Table 2.5), the total bandwidth is large enough. In this case, as α increases, the amount of total bandwidth increases, and the optimal price decreases to maximize the profit.

Figure 2.7 shows how the optimal price changes with various costs and α in the low sensing cost regime. It is clear that price is first a constant and then starts to decrease when α is larger than a threshold. The threshold decreases in the optimal sensing decision of B_s^{L*}: a smaller sensing cost or a higher leasing cost will lead to a higher B_s^{L*} and thus a smaller threshold.

It is interesting to notice that the equilibrium price only changes in a time slot where the sensing realization factor α is large. This means that although operator has the freedom to change the price in every time slot, the actual variation of price is much less frequent. This makes it easier to implement in practice. Figure 2.8 illustrates this with different sensing costs and α realizations. In each time slot, a realization of α distribution is drawn and we can derive equilibrium price from Table 2.5. The left two subfigures correspond to the realizations of α and the corresponding prices with $C_s = 0.48$ and $C_l = 1$. As the sensing cost C_s is quite high in this case, the operator does not rely heavily on sensing. As a result, the variability of α (in the upper subfigure) has very small impact on the equilibrium price (in the lower subfigure). In fact, the price only changes in 11 out 50 time slots, and the maximum amplitude variation is around 10 %. The right two figures correspond to the case where $C_s = 0.35$ and $C_l = 1$. As sensing cost is cheaper in this case, the operator senses more and the impact of α on price is larger. The price changes in 30 out of 50 time slots, and the variation in amplitude can be as large as 30 %.

Observation 4 *The operator will sense the spectrum only if the sensing cost is lower than a threshold. Also, tt will lease additional spectrum only if the spectrum obtained through sensing is below a threshold. Furthermore, it will charge a constant price to the users if the total bandwidth obtained through sensing and leasing does not exceed a threshold.*

Observation 5 *Each user i obtains the same SNR independent of g_i and a payoff linear in g_i.*

Fig. 2.7 Optimal price π^* as a function of C_s, C_l, and α

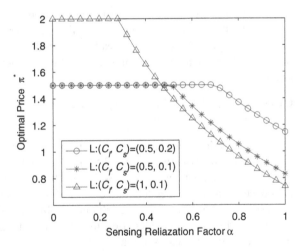

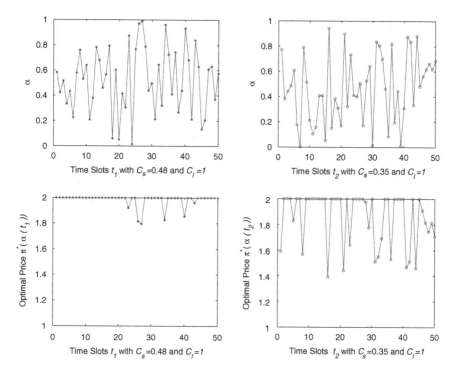

Fig. 2.8 Optimal price π^* over time with different sensing costs and α realizations

Observation 5 shows that users obtains fair and predictable resource allocation at the equilibrium. In fact, a user does not need to know anything about the total number and payoffs of other users in the system. It can simply predict its QoS if it knows the cost structure of the network (C_s and C_l).[11] Such property is highly desirable in practice.

Finally, users achieve the same high SNR at the equilibrium. The SNR value is either $e^{(2+C_l)}$ or $G/(B_s^{L*}\alpha)$, both of which are larger than e^2. This means that the approximation ratio $\ln(\text{SNR}_i)/\ln(1+\text{SNR}_i) > \ln(e^2)/\ln(1+e^2) \approx 94\%$. The ratio can even be close to one if the price π is high.

In Sects. 2.3.1 and 2.3.4, we made the high SNR regime approximation and the uniform distribution assumption of α to obtain closed-form expressions. Next we show that relaxing both assumptions will not change any of the major insights.

---

[11] The analysis of the game, however, does not require the users to know C_s or C_l.

2.4.1 Robustness of the Observations

Theorem 2.4. *Observations 1-5 still hold under the general SNR regime (as in (2.1)) and any general distribution of α.*

Proof We represent a user i's payoff function in the general SNR regime,

$$u_i(\pi, w_i) = w_i \ln\left(1 + \frac{g_i}{w_i}\right) - \pi w_i. \tag{2.15}$$

The optimal demand $w_i^*(\pi)$ that maximizes (2.15) is $w_i^*(\pi) = g_i/Q(\pi)$, where $Q(\pi)$ is the unique positive solution to $F(\pi, Q) := \ln(1 + Q) - \frac{Q}{1+Q} - \pi = 0$. We find the inverse function of $Q(\pi)$ to be $\pi(Q) = \ln(1 + Q) - \frac{Q}{1+Q}$. By applying the implicit function theorem, we can obtain the first-order derivative of function $Q(\pi)$ over π as

$$Q'(\pi) = -\frac{\partial F(\pi, Q)/\partial \pi}{\partial F(\pi, Q)/\partial Q} = \frac{(1 + Q(\pi))^2}{Q(\pi)}, \tag{2.16}$$

which is always positive. Hence, $Q(\pi)$ is increasing in π.

User i's optimal payoff is

$$u_i(\pi, w_i^*(\pi)) = \frac{g_i}{Q(\pi)}[\ln(1 + Q(\pi)) - \pi]. \tag{2.17}$$

As a result, a user's optimal SNR equals $g_i/w_i^*(\pi) = Q(\pi)$ and is *user-independent*. The total demand from all users equals $G/Q(\pi)$, and the operator's investment and pricing problem is

$$R^* = \max_{B_s \geq 0} E_{\alpha \in [0,1]}[\max_{B_l \geq 0} \max_{\pi \geq 0}(\min\left(\pi \frac{G}{Q(\pi)}, \pi(B_l + B_s\alpha)\right) - B_s C_s - B_l C_l)]. \tag{2.18}$$

Define $\widetilde{R^*} = \frac{R^*}{G}$, $\widetilde{B}_l = \frac{B_l}{G}$, and $\widetilde{B}_s = \frac{B_s}{G}$. Then solving (2.18) is equivalent to solving

$$\widetilde{R^*} = \max_{\widetilde{B}_s \geq 0} E_{\alpha \in [0,1]}[\max_{\widetilde{B}_l \geq 0} \max_{\pi \geq 0}(\min\left(\frac{\pi}{Q(\pi)}, \pi(\widetilde{B}_l + \widetilde{B}_s\alpha)\right) - \widetilde{B}_s C_s - \widetilde{B}_l C_l)]. \tag{2.19}$$

In Problem (2.19), it is clear that the operator's optimal decisions on leasing, sensing and pricing do not depend on users' aggregate wireless characteristics. This is true for any continuous distribution of α. And a user's optimal payoff in Eq. (2.17) is linear in g_i since $Q(\pi)$ is independent of users' wireless characteristics. This shows that Observations 1, 2, and 5 hold for the general SNR regime and any general distribution of α. We can also show that Observations 3 and 4 hold in the general case, with a detailed proof in Appendix 2.8.3.

2.5 The Impact of Spectrum Sensing Uncertainty

The key difference between our model and most existing literature e.g., [37, 41, 43, 47, 49, 50] is the possibility of obtaining resource through the cheaper but uncertain approach of spectrum sensing. Here we will elaborate the impact of sensing on the performances of operator and users by comparing with the *baseline case* where sensing is not possible. Note that in the high sensing cost regime it is optimal not to sense, as a result, the performance of the operator and users will be the same as the baseline case. Hence we will focus on the low sensing cost regime in Table 2.5.

Observation 6 *The operator's optimal expected profit always benefits from the availability of spectrum sensing in the low sensing cost regime.*

Figure 2.9 illustrates the normalized optimal expected profit as a function of the sensing cost. We assume leasing cost $C_l = 2$, and thus the low sensing cost regime corresponds to the case where $C_s \in [0.2, 1]$ in the figure. It is clear that sensing achieves a better optimal expected profit in this regime. In fact, sensing leads to 250% increase in profit when $C_s = 0.2$. The benefit decreases as the sensing cost becomes higher. When sensing becomes too expensive, the operator will choose not to sense and thus achieve the same profit as in the baseline case.

Theorem 2.5. *The operator's realized profit (i.e., the profit for a given α) is a strictly increasing function in α in the low sensing cost regime. Furthermore, there exists a threshold $\alpha_{th} \in (0, 1)$ such that the operator's realized profit is larger than the baseline approach if $\alpha > \alpha_{th}$.*

Proof As in Table 2.5, we have two cases in the low sensing cost regime:

- If $\alpha \le Ge^{-(2+C_l)}/B_s^{L*}$, then substituting B_s^{L*} into $R_{II}^{CS1}(B_s, \alpha)$ in Table 2.3 leads to the realized profit

Fig. 2.9 Operator's normalized optimal *expected* profit as a function of C_s and C_l

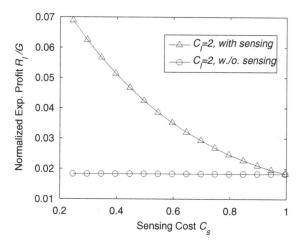

$$R_{II}^{CS1}(\alpha) = Ge^{-(2+C_l)} - B_s^{L*}C_s + B_s^{L*}\alpha C_l,$$

which is strictly and linearly increasing in α.

- If $\alpha \geq Ge^{-(2+C_l)}/B_s^{L*}$, then substituting B_s^{L*} into $R_{II}^{CS2}(B_s, \alpha)$ in Table 2.3 leads to the realized profit

$$R_{II}^{CS2}(\alpha) = B_s^{L*}\alpha \left(\ln \left(\frac{G}{B_s^{L*}\alpha} \right) - 1 \right) - B_s^{L*}C_s.$$

Because the first-order derivative

$$\frac{\partial R_{II}^{CS2}(\alpha)}{\partial \alpha} = B_s^{L*} \left(\ln \left(\frac{G}{B_s^{L*}\alpha} \right) - 2 \right) > 0,$$

as $B_s^{L*} \leq Ge^{-2}$, $R_{II}^{CS2}(\alpha)$ is strictly increasing in α.

We can also verify that $R_{II}^{CS1}(\alpha) = R_{II}^{CS2}(\alpha)$ when $\alpha = Ge^{-(2+C_l)}/B_s^{L*}$. Therefore, the realized profit is a continuous and strictly increasing function of α.

Next we prove the existence of threshold α_{th}. First consider the extreme case $\alpha = 0$. Since the operator obtains no bandwidth through sensing but still incurs some cost, the profit in this case is lower than the baseline case. Furthermore, we can verify that $R_{II}^{CS2}(1) > R_I^H$ in Table 2.5, thus the realized profit at $\alpha = 1$ is always larger than the baseline case. Together with the continuity and strictly increasing nature of the realized profit function, we have proven the existence of threshold of α_{th}.

Figure 2.10 shows the realized profit as a function of α for different costs. The realized profit is increasing in α in both cases. The "crossing" feature of the two increasing curves is because the optimal sensing B_s^* is larger under a cheaper sensing cost ($C_s = 0.5$), which leads to larger realized profit loss (gain, respectively) when $\alpha \to 0$ ($\alpha \to 1$, respectively). This shows the tradeoff between improvement of expected profit and the large variability of the realized profit.

Theorem 2.6. *Users always benefit from the availability of spectrum sensing in the low sensing cost regime.*

Proof In the baseline approach without sensing, the operator always charges the price $1 + C_l$. As shown in Table 2.5, the equilibrium price π^* with sensing is always no larger than $1 + C_l$ for any value of α. Since a user's payoff is strictly decreasing in price, the users always benefit from sensing.

Figure 2.11 shows how a user i's normalized realized payoff u_i^*/g_i changes with α. The payoff linearly increases in α when α becomes larger than a threshold, in which case the equilibrium price becomes lower than $1 + C_l$. A smaller sensing cost C_s leads to more aggressive sensing and thus more benefits to the users.

Fig. 2.10 Operator's normalized optimal *realized* profit as a function of α

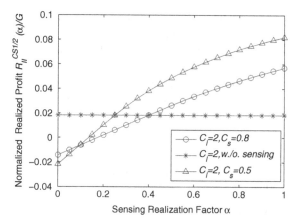

Fig. 2.11 User i's normalized optimal realized payoff as a function of α

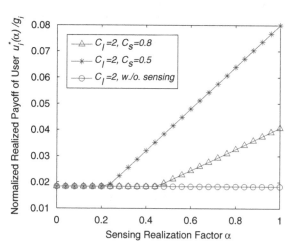

2.6 Learning the Distribution of Sensing Realization Factor α

Our previous analysis assumes that the operator knows the distribution of sensing realization factor α beforehand. When such information is not available, the operator can learn the distribution through machine learning [70]. Next we propose a machine learning algorithm, where the operator uses the sensing realizations of previous time slots to update the distribution of α.

Let us denote the probability density function (pdf) of α as $f(\alpha)$ over the support of $[0, 1]$. Although $f(\alpha)$ is in general continuous, we can approximate it through a proper discretization, i.e., representing the pdf by a probability mass function (pmf) over $N + 1$ equally spaced values (with the first and last values equal to 0 and 1, respectively).

The overall learning process is divided into several learning rounds. Each learning round consists of M time slots. In the kth learning round, the operator builds an empirical distribution of the α distribution, an $N + 1$ long vector $Record_k$, based on the observation of sensing results over the M time slots in this round. At the end of the learning round, the operator updates the α distribution estimation $Distr_{k+1}$ based on the current value of $Distr_k$ and $Record_k$. The machine learning algorithm for updating the distribution of α is shown in Algorithm 1.

Algorithm 1 Machine Learning Algorithm

1: Initiates $Distr_1$ by an arbitrary distribution at $k = 1$.
2: **while** $k = 1$ or $Distr_k \neq Distr_{k-1}$ **do**
3:   Compute B_s^* according to $Distr_k$
4:   Initialize the empirical distribution $Record_k = (0, ..., 0)$ (with $N + 1$ entries)
5:   **for** time slot $m = 1$ to M **do**
6:     Senses the spectrum and records the α realization
7:     Updates $Record_k$ with the current α realization by adding one to the corresponding entry. For example, if $\alpha = 0.36$ and $N = 100$, then the operator increases the 37th entry of $Record_k$ by one.
8:     Compute B_l^* and π^* according to Tables 2.3 and 2.2 in the mth time slot
9:   **end for**
10:   Update $Distr_{k+1} = \beta Distr_k + (1 - \beta) \frac{Record_k}{M}$, where $\beta \in [0, 1]$ is the discount factor
11:   $k := k + 1$
12: **end while**

A proper choice of the learning round length M is important. If M is too large, then the distribution update takes much time. If M is too small, then the operator needs to frequently recompute its sensing decision according to the updated distribution in each round. This increases the computation overhead.

2.6.1 Performance Evaluation of Machine Learning

We evaluate the performance of the proposed machine learning algorithm for updating the α distribution. For the illustration purpose, we assume that α follows a normal distribution with mean $m = 0.5$ and standard deviation $\delta = 0.15$.[12] Notice that the proposed algorithm works for any distribution of α. The operator starts with an initial "guess" $Distr_1$ of uniform distribution. We assume $N = 100$ and $M = 5000$ in the simulation.

Figure 2.12 shows the operator's estimation of the distribution of α over multiple learning rounds with $\beta = 0.8$. It is generated by following Algorithm 1, where at each round the operator's belief of α distribution is updated. The round 1 line corresponds to the uniform distribution of $Distr_1$. After 40 rounds, the operator obtains a pmf that approximations the real normal distribution very well.

---

[12] This choice of m and δ ensures that almost all α realizations fall into the feasible range $[0, 1]$.

Fig. 2.12 Operator's learning of the distribution of α over learning rounds with $\beta = 0.8$

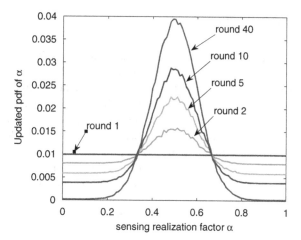

Fig. 2.13 Operator's adaptation of its sensing decision B_s^* over learning rounds with different β values

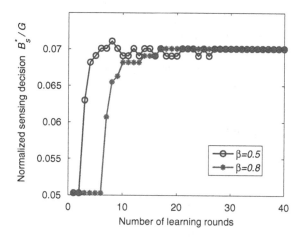

Figure 2.13 illustrates the impact of discount factor β on the convergence of the sensing decision B_s^*. Both curves converge as the number of learning rounds increases. When β is small (e.g., $\beta = 0.5$), the operator's sensing decision converges fast with many fluctuations. When β is large (e.g., $\beta = 0.8$), the operator's sensing decision converges slowly with less fluctuations. The operator needs to trade off convergence speed and fluctuations by choosing the proper β.

In summary, using the proposed machine learning algorithm, the operator can quickly learn the distribution of α, and can make good use of spectrum resource by adapting its sensing decisions dynamically.

2.7 Summary

This chapter represents some encouraging results towards understanding the new
business models, opportunities, and challenges of the emerging cognitive virtual
mobile network operators (C-MVNOs) under supply uncertainty. Here we focus
on studying the trade-off between the cost and uncertainty of spectrum investment
through sensing and leasing. We model the interactions between the operator and the
users by a Stackelberg game, which captures the wireless heterogeneity of users in
terms of maximum transmission power levels and channel gains.

We have discovered several interesting features of the game equilibrium. We show
that the operator's optimal sensing, leasing, and pricing decisions follow nice threshold structures. The availability of sensing always increases the operator's expected
profit, despite that the realized profit in each time slot will have some variations
depending on the sensing result. Moreover, users always benefit in terms of payoffs
when sensing is performed by the operator.

Through the analytical and simulation study of an idealized model in this chapter,
we have obtained various interesting engineering and economical insights into the
operations of C-MVNOs. We hope that this chapter can contribute to the further
understanding of proper network architecture decisions and business models of future
cognitive radio systems.

2.8 Appendix

2.8.1 Proof of Theorem 2.1

Given the total bandwidth $B_l + B_s\alpha$, the objective of Stage III is to solve the optimization problem (2.8), i.e., $\max_{\pi \geq 0} \min(D(\pi), S(\pi))$. First, by examining the derivative of $D(\pi)$, i.e., $\partial D(\pi)/\partial \pi = (1 - \pi)Ge^{-(1+\pi)}$, we can see that the continuous
function $D(\pi)$ is increasing in $\pi \in [0, 1]$ and decreasing in $\pi \in [1, +\infty]$, and $D(\pi)$
is maximized when $\pi = 1$. Since $S(\pi)$ always increases in π and $D(\pi)$ is concave
over $\pi \in [0, 1]$, $S(\pi)$ intersects with $D(\pi)$ if and only if $\frac{\partial D(\pi)}{\partial \pi} > \frac{\partial S(\pi)}{\partial \pi}$ at $\pi = 0$,
i.e., $B_l + B_s\alpha < Ge^{-1}$.

Next we divide our discussion into the intersection case and the non-intersection
case:

1. Given $B_l + B_s\alpha \leq Ge^{-1}$, $S(\pi)$ intersects with $D(\pi)$. By solving equation
   $S(\pi) = D(\pi)$ the intersection point is $\pi = \ln\left(\frac{G}{B_l + B_s\alpha}\right) - 1$. There are two
   subcases:

   - when $B_l + B_s\alpha \leq Ge^{-2}$, $S(\pi)$ intersects with $D(\pi)$, and $\min(D(\pi), S(\pi))$ is
     maximized at the intersection point, i.e., $\pi^* = \ln\left(\frac{G}{B_l + B_s\alpha}\right) - 1$. (See $S_3(\pi)$
     in Fig. 2.3.)

- when $B_l + B_s\alpha \geq Ge^{-2}$, $S(\pi)$ intersects with $D(\pi)$, and $\min(D(\pi), S(\pi))$ is maximized at the maximum value of $D(\pi)$, i.e., $\pi^* = 1$. (See $S_2(\pi)$ in Fig. 2.3.)

2. Given $B_l + B_s\alpha \geq Ge^{-1}$, $S(\pi)$ doesn't intersect with $D(\pi)$. Then $\min(D(\pi), S(\pi))$ is maximized at the maximum value of $D(\pi)$, i.e., $\pi^* = 1$. (See $S_1(\pi)$ in Fig. 2.3.) ∎

2.8.2 Proof of Theorem 2.2

Given the sensing result $B_s\alpha$, the objective of Stage II is to solve the decomposed two subproblems (2.10) and (2.11), and select the best one with better optimal performance. Since $R_{III}^{ES}(B_s, \alpha, B_l)$ in subproblem (2.10) is linearly decreasing in B_l, its optimal solution always lies at the lower boundary of the feasible set (i.e., $B_l^* = \max\{Ge^{-2} - B_s\alpha, 0\}$). We compare the optimal profits of two subproblems (i.e., $R_{II}^{ES}(B_s, \alpha)$ and $R_{II}^{CS}(B_s, \alpha)$) for different sensing results:

1. Given $B_s\alpha > Ge^{-2}$, the obtained bandwidth after Stage I is already in excessive supply regime. Thus it is optimal not to lease for subproblem (2.10) (i.e., $B_l^{ES3} = 0$ of case (ES3) in Table 2.3).

2. Given $0 \leq B_s\alpha \leq Ge^{-2}$, the optimal leasing decision for subproblem (2.11) is $B_l^* = Ge^{-2} - B_s\alpha$ and we have $R_{III}^{ES}(B_s, \alpha, B_l) = R_{III}^{CS}(B_s, \alpha, B_l)$ when $B_l = Ge^{-2} - B_s\alpha$, thus the optimal objective value of (2.10) is always no larger than that of (2.11) and it is enough to consider the conservative supply regime only. Since

$$\frac{\partial^2 R_{III}^{CS}(B_s, \alpha, B_l)}{\partial B_l^2} = -\frac{1}{B_l + B_s\alpha} < 0,$$

$R_{III}^{CS}(B_s, \alpha, B_l)$ is concave in $0 \leq B_l \leq Ge^{-2} - B_s\alpha$. Thus it is enough to examine the first-order condition

$$\frac{\partial R_{III}^{CS}(B_s, \alpha, B_l)}{\partial B_l} = \ln\left(\frac{G}{B_l + B_s\alpha}\right) - 2 - C_l = 0,$$

and the boundary condition $0 \leq B_l \leq Ge^{-2} - B_s\alpha$. This results in optimal leasing decision $B_l^* = \max(Ge^{-(2+C_l)} - B_s\alpha, 0)$ and leads to $B_l^{CS1} = Ge^{-(2+C_l)} - B_s\alpha$ and $B_l^{CS2} = 0$ of cases (CS1) and (CS2) in Table 2.3.

By substituting B_l^{CS1} and B_l^{CS2} into $R_{III}^{CS}(B_s, \alpha, B_l)$ in Table 2.2, we derive the corresponding optimal profits $R_{II}^{CS1}(B_s, \alpha)$ and $R_{II}^{CS2}(B_s, \alpha)$ in Table 2.3. $R_{II}^{ES3}(B_s, \alpha)$ can also be obtained by substituting B_l^{ES3} into $R_{III}^{ES}(B_s, \alpha, B_l)$. ∎

2.8.3 Supplementary Proof of Theorem 2.4

In this section, we prove that Observations 3 and 4 hold for the genera case (i.e., the general SNR regime and a general distributions of α). We first show that Observation 4 holds for the general case.

2.8.3.1 Threshold Structure of Sensing

It is not difficult to show that if the sensing cost is much larger than the leasing cost, the operator has no incentive to sense but will directly lease. Thus the threshold structure on the sensing decision in Stage I still holds for the general case. We ignore the details due to space limitations.

2.8.3.2 Threshold Structure of Leasing

Next we show the threshold structure on leasing in Stage II also holds. Similar as in the proof of Theorem 2.1, we define $D(\pi) = \pi \frac{G}{Q(\pi)}$ and $S(\pi) = \pi(B_s\alpha + B_l)$.

- We first show that $D(\pi)$ is increasing when $\pi \in [0, 0.468]$ and decreasing when $\pi \in [0.468, +\infty)$. To see this, we take the first-order derivative of $D(\pi)$ over π,

$$D'(\pi) = \frac{2Q(\pi)^2 + Q(\pi) - (1 + Q(\pi))^2 \ln(1 + Q(\pi))}{Q(\pi)^3},$$

which is positive when $Q(\pi) \in [0, 2.163)$ and negative when $Q(\pi) \in [2.163, +\infty)$. Since Eq. (2.16) shows that $Q(\pi)$ is increasing in π and $\pi(Q)|_{Q=2.163} = 0.468$, as a result $D(\pi)$ is increasing in $\pi \in [0, 0.468]$ and decreasing in $\pi \in [0.468, +\infty)$. In other words, $D(\pi)$ is maximized at $\pi = 0.468$.
- Next we derive the operator's optimal pricing decision in Stage III. Figure 2.15 shows two possible intersection cases of $S(\pi)$ and $D(\pi)$. B_{th1} is defined as the total bandwidth obtained in Stages I and II (i.e., $B_s\alpha + B_l$) such that $S(\pi)$ intersects with $D(\pi)$ at $\pi = 0.468$. Here is how the optimal pricing is determined:

  - If $B_s\alpha + B_l \geq B_{th1}$ (e.g., $S_1(\pi)$ in Fig. 2.15), the optimal price is $\pi^* = 0.468$. The total supply is no smaller (and often exceeds) the total demand.
  - If $B_s\alpha + B_l < B_{th1}$ (e.g., $S_2(\pi)$ in Fig. 2.15), the optimal price occurs at the unique intersection point of $S(\pi)$ and $D(\pi)$ (where $D(\pi)$ has a negative first-order derivative). The total supply equals total demand.

- Now we are ready to show the threshold structure of the leasing decision.

  - If the sensing result from Stage I satisfies $B_s\alpha \geq B_{th1}$, then the operator will not lease. This is because leasing will only increase the total cost without increasing

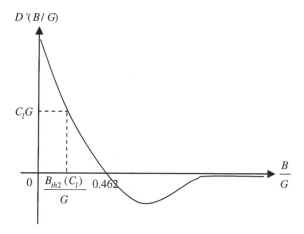

Fig. 2.14 The relation between the normalized total bandwidth B/G and the derivative of the revenue $D'(B/G)$

the revenue, since the optimal price is fixed at $\pi^* = 0.468$ and thus revenue is also fixed at $D(\pi^*)$.

– Let us focus on the case where the sensing result from Stage I satisfies $B_s\alpha < B_{th1}$. Let us define $B = B_s\alpha + B_l$, then we have $B = G/Q(\pi)$ and $\pi = \ln(1 + G/B) - G/(G + B)$. This enables us to rewrite $D(\pi)$ as a function of total resource B only,

$$D(B) = B\left[\ln\left(1 + \frac{G}{B}\right) - \frac{G}{G + B}\right].$$

The first-order derivative of $D(B)$ is

$$D'(B) = \ln\left(1 + \frac{1}{B/G}\right) - \frac{1}{1 + B/G} - \frac{1}{(1 + B/G)^2}, \qquad (2.20)$$

which denotes the increase of revenue $D(B)$ due to unit increase in bandwidth B. Since obtaining each unit bandwidth has a cost of C_l in Stage II, the operator will only lease positive amount of bandwidth if and only if $D'(B_s\alpha) > C_l$. To facilitate the discussions, we will plot the function of $D'(B/G)$ in Fig. 2.14, with the understanding that $D'(B/G) = D'(B)G$. The intersection point of $B/G = 0.462$ in Fig. 2.14 corresponds to the point of $\pi = 0.468$ in Fig. 2.15. The positive part of $D'(B)$ on the left side of $B/G = 0.462$ in Fig. 2.14 corresponds to the part of $D(\pi)$ with a negative first-order derivative in Fig. 2.15.

For any value C_l, Fig. 2.14 shows that there exists a unique threshold $B_{th2}(C_l)$ such that $D'(B_{th2}(C_l)/G) = C_lG$, i.e., $D'(B_{th2}(C_l)) = C_l$. Then the optimal leasing amount will be $B_{th2}(C_l) - B_s\alpha$ if the bandwidth obtained from sensing $B_s\alpha$ is less than $B_{th2}(C_l)$, otherwise it will be zero.

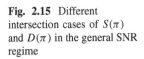

Fig. 2.15 Different intersection cases of $S(\pi)$ and $D(\pi)$ in the general SNR regime

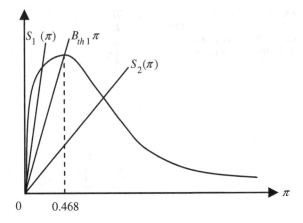

2.8.3.3 Threshold Structure of Pricing and Observation 3

Based on the proofs above, we show that Observation 3 also holds for the general case as follows. Let us denote the optimal sensing decision as B_s^*, and consider two sensing realizations α_1 and α_2 in time slots 1 and 2, respectively. Without loss of generality, we assume that $\alpha_1 < \alpha_2$.

- If $B_s^* \alpha_2 \geq B_{th1}$, then the optimal price in time slot 2 is $\pi^* = 0.468$ (see Fig. 2.15). The optimal price in time slot 1 is always no smaller than 0.468.
- If $B_s^* \alpha_1 < B_s^* \alpha_2 < B_{th1}$, then we need to consider three subcases:

  - If $B_s^* \alpha_1 < B_s^* \alpha_2 \leq B_{th2}(C_l)$, then the operator will lease up to the threshold in both time slots, i.e., $B_l^* = B_{th2}(C_l) - B_s^* \alpha_1$ in time slot 1 and $B_l^* = B_{th2}(C_l) - B_s^* \alpha_2$ in time slot 2. Then optimal prices in both time slots are the same.
  - If $B_s^* \alpha_1 \leq B_{th2}(C_l) < B_s^* \alpha_2$, then the operator will lease $B_l^* = B_{th2}(C_l) - B_s^* \alpha_1$ in time slot 1 and will not lease in time slot 2. Thus the total bandwidth in time slot 1 is smaller than that of time slot 2, and the optimal price in time slot 1 is larger.
  - If $B_{th2}(C_l) \leq B_s^* \alpha_1 < B_s^* \alpha_2$, then the operator in both time slots will not lease and total bandwidth in time slot 1 is smaller, and the optimal price in time slot 1 is larger.

To summarize, the optimal price π^* in Stage III is non-increasing in α. And the operator will charge a constant price ($\pi^* = 0.468$) to the users as long as the total bandwidth obtained through sensing and leasing does not exceed the threshold $B_{th2}(C_l)$.

∎

Chapter 3
Secondary Spectrum Market Under Operator Competition

Abstract This chapter presents a comprehensive analytical study of two competitive secondary operators' investment (i.e., spectrum leasing) and pricing strategies, taking into account operators' heterogeneity in leasing costs and users' heterogeneity in transmission power and channel conditions. We model the interactions between operators and users as a three-stage dynamic game, where operators simultaneously make spectrum leasing decisions in Stage I, and pricing decisions in Stages II, and then users make purchase decisions in Stage III. We show that both operators' investment and pricing equilibrium decisions process interesting threshold properties. Moreover, two operators always choose the same equilibrium price despite their heterogeneity in leasing costs. Each user fairly achieves the same service quality in terms of signal-to-noise-ratio (SNR) at the equilibrium, and the obtained predictable payoff is linear in its transmission power and channel gain. We also the maximum loss of total profit due to operators' competition is no larger than 25 %. The users, however, always benefit from operators' competition in terms of their payoffs.

3.1 Background

Recall that in Chap. 2, we only study one secondary operator's economic decisions and our focus is how the supply uncertainty in spectrum sensing affect the operator's investment and pricing. However, as there are many operators in the global spectrum market,[1] some operators are competing for the same local market in terms of spectrum acquisition and service pricing (e.g., Virgin Mobile USA and Simple Mobile compete in the same California market).

In this chapter, we study the competition between secondary operators in spectrum acquisition and pricing to serve a common pool of secondary users. To abstract

---

[1] Started from late 1990s, there are over 400 mobile virtual network operators owned by over 360 companies worldwide as of February 2009 [72]. According to Visiongain Report [6], the global MVNO market will be worth $40.55 billion by 2016.

L. Duan et al., *Cognitive Virtual Network Operator Games*,
SpringerBriefs in Computer Science, DOI: 10.1007/978-1-4614-8890-3_3,
© The Author(s) 2013

the interactions among operators, we focus on two operator case (i.e., duopoly) and will study multiple operator case (i.e., oligopoly) in our future work. As the operator competition in the current spectrum market only focuses on spectrum leasing and has not introduce spectrum sensing yet, we only consider the operators' spectrum leasing approach to acquire resource. The operators will dynamically lease spectrum from primary operators, and then compete to sell the resource to the secondary users to maximize their individual profits. We would like to understand *how the operators make the equilibrium investment (leasing) and pricing (selling) decisions, considering operators' heterogeneity in leasing costs and wireless users' heterogeneity in transmission power and channel conditions.*

We adopt a three-stage dynamic game model to study the (secondary) operators' investment and pricing decisions as well as the interactions between the operators and the (secondary) users. In Stage I, the two operators simultaneously lease spectrum (bandwidth) from the primary operators with different leasing costs. In Stage II, the two operators simultaneously announce their spectrum retail prices to the users. In Stage III, each user determines how much resource to purchase from which operator. Each operator wants to maximize its profit, which is the difference between the revenue collected from its users and the cost paid to the primary operator.

Key results and contributions of this chapter include:

- *An appropriate wireless spectrum sharing model*: We assume that heterogeneous users share the spectrum using orthogonal frequency division multiplexing (OFDM) technology. Then a user's achievable rate and thus its spectrum demand depend on its allocated bandwidth, maximum transmission power, and channel condition. This model is more suitable to our problem than the generic economic models used in related literature [37, 47, 49, 50]. It can also provide more engineering insights on how different wireless network parameters in the spectrum sharing model (e.g., users' various wireless characteristics) impact the operators' leasing and pricing decisions.
- *Symmetric pricing structure*: We show the two operators always choose the same equilibrium price, even when they have different leasing costs and make different investment decisions. Moreover, this price is independent of users' transmission powers and channel conditions.[2]
- *Threshold structures of investment and pricing equilibrium*: We show that both operators' investment and pricing equilibrium decisions process interesting threshold properties. For example, when the two operators' leasing costs are close, both operators will lease positive spectrum. Otherwise, one operator will choose not to lease and the other operator becomes the monopolist. For pricing, a positive pure strategy equilibrium exists only when the total spectrum investment of both operators is less than a threshold.

---

[2] Such independency is good for the development of spectrum market, since a user does not need to worry about how variations of user population and wireless characteristics change its performance in spectrum trading.

- *Fair service quality achieved by users*: We show that each user achieves the same signal-to-noise (SNR) that is independent of the users' population and wireless characteristics.
- *Impact of competition*: We show that the operators' competition leads to a maximum 25 % loss of their total profit compared with a coordinated case. The users, however, always benefit from the operators' competition by achieving better payoffs.

Next we briefly discuss the related literature. In Sect. 3.2, we describe the network model and game formulation. In Sect. 3.3, we analyze the dynamic game through backward induction and calculate the duopoly leasing/pricing equilibrium. We discuss various insights obtained from the equilibrium analysis in Sect. 3.4. In Sect. 3.6, we show the impact of duopoly competition on the total operators' profit and the users' payoffs. We conclude in Sect. 3.7 together with some future research directions.

3.2 Network and Game Model

We consider two operators ($i, j \in \{1, 2\}$ and $i \neq j$) and a set $\mathcal{K} = \{1, \ldots, K\}$ of users in an ad hoc network as shown in Fig. 3.1. The operators obtain wireless

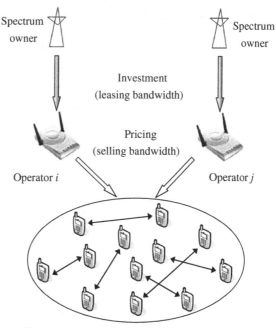

Fig. 3.1 Network model for the secondary network operators

Secondary users (transmitter-receiver pairs)

spectrum from different primary operators with different leasing costs, and compete to serve the same set $\mathcal{K}$ of users. Each user has a transmitter-receiver pair. We assume that users are equipped with software defined radios and can transmit in a wide range of frequencies as instructed by the operators, but do not have the capability of spectrum sensing in cognitive radios.[3] Such a network structure puts most of the implementation complexity for dynamic spectrum leasing and allocation on the operators, and thus is easier to implement than a "full" cognitive radio network especially for a large number of users. A user may switch among different operators' services (e.g. WiMAX, 3G) depending on operators' prices. It is important to study the competition among multiple operators as operators are normally not cooperative.

The interactions between the two operators and the users can be modeled as a *three-stage dynamic game*, as shown in Fig. 3.2. Operators i and j first simultaneously determine their leasing bandwidths in Stage I, and then simultaneously announce the prices to the users in Stage II. Finally, each user chooses to purchase bandwidth from *only one operator* to maximize its payoff in Stage III.

The key notations of the chapter are listed in Table 3.1. Some are explained as follows.

- *Leasing decisions B_i and B_j*: leasing bandwidths of operators i and j in Stage I, respectively.
- *Costs C_i and C_j*: the fixed positive leasing costs per unit bandwidth for operators i and j, respectively. These costs are determined by the negotiation between the operators and their own spectrum suppliers.
- *Pricing decisions p_i and p_j*: prices per unit bandwidth charged by operators i and j to the users in Stage II, respectively.
- *The User k's demand w_{ki} or w_{kj}*: the bandwidth demand of a user $k \in \mathcal{K}$ from operator i or j. A user can only purchase bandwidth from one operator.

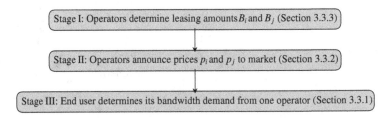

Fig. 3.2 Three-stage dynamic game: the duopoly's leasing and pricing, and the users' resource allocation

---

[3] Spectrum sensing is the most important functionality of cognitive radios, which enables users to actively monitor the external radio environments to communicate efficiently without interfering primary users. The capability of spectrum sensing includes comprehensive monitoring of frequency spectrum, user behavior, and network state over time.

Table 3.1 Key notations

Notations	Physical meaning
B_i, B_j	Leasing bandwidths of operators i and j
C_i, C_j	Costs per unit bandwidth paid by operators i and j
p_i, p_j	Prices per unit bandwidth announced by operators i and j
$\mathcal{K} = \{1, \ldots, K\}$	Set of the users in the network
$P_k^{\max}$	User k's maximum transmission power
h_k	User k's channel gain between its transceiver
n_0	Noise power per unit bandwidth
$g_k = P_k^{\max} h_k / n_0$	User k's wireless characteristic
$G = \sum_{k \in \mathcal{K}} g_k$	The users' aggregate wireless characteristics
w_{ki}, w_{kj}	User k's bandwidth demand from operator i or j
r_k	User k's data rate
$\mathcal{K}_i^P, \mathcal{K}_j^P$	Preferred user sets of operators i and j
D_i, D_j	Preferred demands of operators i and j
$\mathcal{K}_i^R, \mathcal{K}_j^R$	Realized user sets of operators i and j
Q_i, Q_j	Realized demands of operators i and j
R_i, R_j	Revenues of operators i and j
π_i, π_j	Profits of operators i and j
T_π	Total profit of both operators

3.2.1 Users' and Operators' Models

OFDM has been proposed as a promising physical layer choice for dynamic spectrum sharing [81, 83]. We assume that the users share the spectrum using OFDM to avoid mutual interferences. The main analysis in this chapter assumes that users are located close-by, and thus no two users will transmit over the same channel (also called subcarriers in the OFDM literatures [84, 85]). We also relax this assumption in our online technical report [90] and show that our results can be extended to the case with spectrum spatial reuse.

If a user $k \in \mathcal{K}$ obtains bandwidth w_{ki} from operator i, then it achieves a data rate (in nats) of [86]

$$r_k(w_{ki}) = w_{ki} \ln \left(1 + \frac{P_k^{\max} h_k}{n_0 w_{ki}} \right), \tag{3.1}$$

where $P_k^{\max}$ is user k's maximum transmission power, n_0 is the noise power density, h_k is the channel gain between user k's transmitter and receiver. The channel gain h_k is independent of the operator, as the operator only sells bandwidth and does not provide a physical infrastructure.[4] Here we assume that user k spreads its power $P_k^{\max}$ across the entire allocated bandwidth w_{ki}.

---

[4] We also assume that the channel condition is independent of transmission frequencies, such as in the current 802.11d/e standard [87] where the channels are formed by interleaving over the tones. In other words, each user experiences a flat fading over the entire spectrum.

To simplify later discussions, we let

$$g_k = P_k^{\max} h_k / n_0,$$

thus g_k / w_{ki} is the user k's SNR. The rate in (3.1) is calculated based on the Shannon capacity.

To obtain closed-form solutions, we first focus on the high SNR regime where SNR $\gg 1$. This will be the case where a user has limited choices of modulation and coding schemes, and thus can not decode a transmission if the SNR is below some threshold. In the high SNR regime, the rate in (3.1) can be approximated as

$$r_k(w_{ki}) = w_{ki} \ln \left(\frac{g_k}{w_{ki}} \right). \tag{3.2}$$

Although the analytical solutions in Sect. 3.3 are derived based on (3.2), we will show later in Sect. 3.5 that *all major engineering insights remain unchanged in the general SNR regime.*

If a user k purchases bandwidth w_{ki} from operator i, it receives a *payoff* of

$$u_k(p_i, w_{ki}) = w_{ki} \ln \left(\frac{g_k}{w_{ki}} \right) - p_i w_{ki}, \tag{3.3}$$

which is the difference between the data rate and the payment. The payment is proportional to price p_i announced by operator i. This linear pricing scheme has been widely used in the literature [88, 89].

For an operator i, its profit is the difference between the revenue and the total cost, i.e.,

$$\pi_i(B_i, B_j, p_i, p_j) = p_i Q_i(B_i, B_j, p_i, p_j) - B_i C_i, \tag{3.4}$$

where $Q_i(B_i, B_j, p_i, p_j)$ and $Q_j(B_i, B_j, p_i, p_j)$ are realized demands of operators i and j. The concept of realized demand will be defined later in Definition 3.4.

3.3 Backward Induction of the Three-Stage Game

A common approach of analyzing dynamic game is backward induction to find the subgame perfect equilibrium (SPE) [82]. Subgame perfect equilibrium (or simply, equilibrium) represents a Nash equilibrium of every subgame of the original game. In this chapter, we start with Stage III and analyze the users' behaviors given the operators' investment and pricing decisions. Then we look at Stage II and analyze how operators make the pricing decisions taking the users' demands in Stage III into consideration. Finally, we look at the operators' leasing decisions in Stage I knowing the results in Stages II and III. Throughout the chapter, we will use "bandwidth", "spectrum", and "resource" interchangeably.

In the following analysis, we only focus on pure strategy SPE and rule out mixed SPE in the multi-stage game.[5] Such a methodology has been widely used in the literature [96, 97]. Following the definition in Ref. [96], we use *conditionally* SPE to denote an SPE with pure strategies only, where the network's pure strategies constitute a Nash equilibrium in every subgame. The concept of conditionally SPE is motivated by the concept of SPE but rules out mixed strategies. In Sect. 3.3.2, we will show that a conditionally SPE will not include any investment decisions (B_i, B_j) in the medium investment regime in Stage I. Otherwise there is no pure strategy Nash equilibrium for pricing in Stage II, and it will not be a conditionally SPE.[6]

Following very similar statements in Ref. [96], we list several reasons to focus on conditionally SPE in this chapter without considering mixed strategies.

- First, we want to emphasize the result that a pure strategy pricing equilibrium may not exist in Stage II, as this result highlights the very important Edgeworth paradox for the medium investment regime (which will be introduced in Sect. 3.3.2). Such result reveals the special structure of our problem and leads to important engineering insights for practical network design.
- Second, a standard criticism of mixed strategy equilibrium is that they impose very large informational burdens on users [82]. If operators choose prices according to mixed strategies, users need to consider price distributions (from which the final prices will be drawn by operators) when they choose which operator to purchase from. When the operators' leasing costs change over time, the leasing amounts and the corresponding mixed pricing strategies can also be time-varying. Given all these complexities, it is unlikely that end users will have the computational capacities and willingness to calculate the "equilibrium choices" in real spectrum market. In other words, the analysis results when allowing mixed strategies may not be very relevant for engineering practice.
- Third, two operators need to run the randomization procedure in the pricing stage of each time slot if they adopt mixed pricing strategies. However, such randomization over time may be too complicated to implement in practice in a short time scale [94].

In the following analysis, we derive the conditionally SPE, which is also referred to as equilibrium for simplicity.

3.3.1 Spectrum Allocation in Stage III

In Stage III, each user needs to make the following two decisions based on the prices p_i and p_j announced by the operators in Stage II:

---

[5] For interested readers, we have provided some preliminary analysis of mixed strategy SPE in Ref. [90].

[6] If we do not focus on the concept of conditionally SPE, there may be an SPE with mixed strategies. For example, in the pricing subgame in Stage II, mixed pricing strategy Nash equilibrium can exist in the medium investment regime, which is supported by our analysis in Ref. [90, 95].

1. Which operator to choose?
2. How much to purchase?

If a user $k \in \mathcal{K}$ obtains bandwidth w_{ki} from operator i, then its payoff $u_k(p_i, w_{ki})$ is given in (3.3). Since this payoff is concave in w_{ki}, the unique *demand* that maximizes the payoff is

$$w_{ki}^*(p_i) = \arg \max_{w_{ki} \geq 0} u_k(p_i, w_{ki}) = g_k e^{-(1+p_i)}. \qquad (3.5)$$

Demand $w_{ki}^*(p_i)$ is always positive, linear in g_k, and decreasing in price p_i. Since g_k is linear in channel gain h_k and transmission power $P_k^{\max}$, then a user with a better channel condition or a larger transmission power has a larger demand. It is clear that $w_{ki}^*(p_i)$ is upper-bounded by $g_k e^{-1}$ for any choice of price $p_i \geq 0$. In other words, even if operator i announces a zero price, user k will not purchase infinite amount of resource since it can not decode the transmission if $\text{SNR}_k = g_k/w_{ki}$ is low.

Equation (3.5) shows that every user purchasing bandwidth from operator i obtains the same SNR

$$\text{SNR}_k = \frac{g_k}{w_{ki}^*(p_i)} = e^{1+p_i},$$

and obtains a payoff linear in g_k

$$u_k(p_i, w_{ki}^*(p_i)) = g_k e^{-(1+p_i)}.$$

3.3.1.1 Which Operator to Choose?

Next we explain how each user decides which operator to purchase from. The following definitions help the discussions.

Definition 3.1 The *Preferred User Set* $\mathcal{K}_i^P$ includes the users who prefer to purchase from operator i.

Definition 3.2 The *Preferred Demand* D_i is the total demand from users in the preferred user set $\mathcal{K}_i^P$, i.e.,

$$D_i(p_i, p_j) = \sum_{k \in \mathcal{K}_i^P(p_i, p_j)} g_k e^{-(1+p_i)}. \qquad (3.6)$$

The notations in (3.6) imply that both set $\mathcal{K}_i^P$ and demand D_i only depend on prices (p_i, p_j) in Stage II and are independent of operators' leasing decisions (B_i, B_j) in Stage I. Such dependance can be discussed in two cases:

1. *Different Prices* $(p_i < p_j)$: every user $k \in \mathcal{K}$ *prefers* to purchase from operator i since

$$u_k(p_i, w_{ki}^*(p_i)) > u_k(p_j, w_{kj}^*(p_j)).$$

We have $\mathcal{K}_i^P = \mathcal{K}$ and $\mathcal{K}_j^P = \emptyset$, and

$$D_i(p_i, p_j) = Ge^{-(1+p_i)} \text{ and } D_j(p_i, p_j) = 0,$$

where $G = \sum_{k \in \mathcal{K}} g_k$ represents the aggregate wireless characteristics of the users. This notation will be used heavily later in the chapter.

2. *Same Prices* ($p_i = p_j = p$): every user $k \in \mathcal{K}$ is indifferent between the operators and randomly chooses one with equal probability. In this case,

$$D_i(p, p) = D_j(p, p) = Ge^{-(1+p)}/2.$$

Now let us discuss how much demand an operator can actually satisfy, which depends on the bandwidth investment decisions (B_i, B_j) in Stage I. It is useful to define the following terms.

Definition 3.3 The *Realized User Set* $\mathcal{K}_i^R$ includes the users whose demands are satisfied by operator i.

Definition 3.4 The *Realized Demand* Q_i is the total demand of users in the Realized User Set $\mathcal{K}_i^R$, i.e.,

$$Q_i\left(B_i, B_j, p_i, p_j\right) = \sum_{k \in \mathcal{K}_i^R\left(B_i, B_j, p_i, p_j\right)} g_k e^{-(1+p_i)}.$$

Notice that both $\mathcal{K}_i^R$ and Q_i depend on prices (p_i, p_j) in Stage II and leasing decisions (B_i, B_j) in Stage I. Calculating the Realized Demands also requires considering two different pricing cases.

1. *Different prices* ($p_i < p_j$): The Preferred Demands are $D_i(p_i, p_j) = Ge^{-(1+p_i)}$ and $D_j(p_i, p_j) = 0$.

   - *If Operator i has enough resource* $\left(i.e., B_i \geq D_i\left(p_i, p_j\right)\right)$: all Preferred Demand will be satisfied by operator i. The Realized Demands are

   $$Q_i = \min(B_i, D_i(p_i, p_j)) = Ge^{-(1+p_i)},$$
   $$Q_j = 0.$$

   - *If Operator i has limited resource* $\left(i.e., B_i < D_i\left(p_i, p_j\right)\right)$: since operator i cannot satisfy the Preferred Demand, some demand will be satisfied by operator j if it has enough resource. Since the realized demand

   $$Q_i(B_i, B_j, p_i, p_j) = B_i = \sum_{k \in \mathcal{K}_i^R} g_k e^{-(1+p_i)},$$

then $\sum_{k \in \mathcal{K}_i^R} g_k = B_i e^{1+p_i}$.[7] The remaining users want to purchase bandwidth from operator j with a total demand of $\frac{G - B_i e^{1+p_i}}{e^{1+p_j}}$. Thus the Realized Demands are

$$Q_i = \min(B_i, D_i(p_i, p_j)) = B_i,$$
$$Q_j = \min\left(B_j, \frac{G - B_i e^{1+p_i}}{e^{1+p_j}}\right).$$

2. *Same prices* ($p_i = p_j = p$): both operators will attract the same Preferred Demand $Ge^{-(1+p)}/2$. The Realized Demands are

$$Q_i = \min\left(B_i, D_i(p, p) + \max\left(D_j(p, p) - B_j, 0\right)\right)$$
$$= \min\left(B_i, \frac{G}{2e^{1+p}} + \max\left(\frac{G}{2e^{1+p}} - B_j, 0\right)\right),$$
$$Q_j = \min\left(B_j, D_j(p, p) + \max\left(D_i(p, p) - B_i, 0\right)\right)$$
$$= \min\left(B_j, \frac{G}{2e^{1+p}} + \max\left(\frac{G}{2e^{1+p}} - B_i, 0\right)\right).$$

3.3.2 Operators' Pricing Competition in Stage II

In Stage II, the two operators simultaneously determine their prices (p_i, p_j) considering the users' preferred demands in Stage III, given the investment decisions (B_i, B_j) in Stage I.

An operator i's profit is defined earlier in (3.4). Since the payment $B_i C_i$ is fixed at this stage, operator i's profit maximization problem is equivalent of maximization of its revenue $p_i Q_i$. Note that users' total demand Q_i to operator i depends on the received power of each user (product of its transmission power and channel gain). We assume that an operator i knows users' transmission powers and channel conditions. This can be achieved similarly as in today's cellular networks, where users need to register with the operator when they enter the network and frequently feedback the channel conditions. Thus we assume that an operator knows the user population and user demand.

Game 1 (Pricing Game) *The competition between the two operators in Stage II can be modeled as the following game:*

- *Players: two operators i and j.*

---

[7] In this chapter, we consider a large number of users and each user is non-atomic (infinitesimal). Thus an individual user's demand is infinitesimal to an operator's supply and we can claim equality holds for $Q_i = B_i$.

- *Strategy space: operator i can choose price p_i from the feasible set $\mathscr{P}_i = [0, \infty)$. Similarly for operator j.*
- *Payoff function: operator i wants to maximize the revenue $p_i Q_i(B_i, B_j, p_i, p_j)$. Similarly for operator j.*

At an equilibrium of the pricing game, (p_i^*, p_j^*), each operator maximizes its payoff assuming that the other operator chooses the equilibrium price, i.e.,

$$p_i^* = \arg \max_{p_i \in \mathscr{P}_i} p_i Q_i(B_i, B_j, p_i, p_j^*), \quad i = 1, 2, i \neq j.$$

In other words, no operator wants to unilaterally change its pricing decision at an equilibrium.

Next we will investigate the existence and uniqueness of the pricing equilibrium. First, we show that it is sufficient to only consider symmetric pricing equilibrium for Game 1.

Proposition 3.1 *Assume both operators lease positive bandwidth in Stage I, i.e., $\min(B_i, B_j) > 0$. If pricing equilibrium exists, it must be symmetric $p_i^* = p_j^*$.*

The proof of Proposition 3.1 is given in Ref. [90]. The intuition is that no operator will announce a price higher than its competitor to avoid losing its Preferred Demand. This property significantly simplifies the search for all possible equilibria.

Next we show that the symmetric pricing equilibrium is a function of (B_i, B_j) as shown in Fig. 3.3.

Theorem 3.1 *The equilibria of the pricing game are as follows.*

- Low Investment Regime: $(B_i + B_j \leq Ge^{-2}$ as in region (L) of Fig. 3.3): there exists a unique nonzero pricing equilibrium

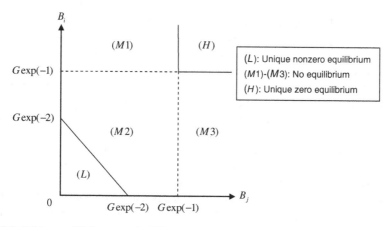

Fig. 3.3 Pricing equilibrium types in different (B_i, B_j)

$$p_i^*(B_i, B_j) = p_j^*(B_i, B_j) = \ln\left(\frac{G}{B_i + B_j}\right) - 1. \qquad (3.7)$$

The operators' profits in Stage II are

$$\pi_{II,\,i}(B_i, B_j) = B_i\left(\ln\left(\frac{G}{B_i + B_j}\right) - 1 - C_i\right), \qquad (3.8)$$

$$\pi_{II,\,j}(B_i, B_j) = B_j\left(\ln\left(\frac{G}{B_i + B_j}\right) - 1 - C_j\right). \qquad (3.9)$$

- Medium Investment Regime ($B_i + B_j > Ge^{-2}$ and $\min(B_i, B_j) < Ge^{-1}$ as in regions (M1)–(M3) of Fig. 3.3): *there is no pricing equilibrium.*
- High Investment Regime ($\min(B_i, B_j) \geq Ge^{-1}$ as in region (H) of Fig. 3.3): *there exists a unique zero pricing equilibrium*

$$p_i^*(B_i, B_j) = p_j^*(B_i, B_j) = 0,$$

and the operators' profits are negative for any positive values of B_i and B_j.

Proof of Theorem 3.1 is given in Appendix 3.8.1. Intuitively, higher investments in Stage I will lead to lower equilibrium prices in Stage II. Theorem 3.1 shows that the only interesting case is the low investment regime where both operators' total investment is no larger than Ge^{-2}, in which case there exists a unique positive symmetric pricing equilibrium. Notice that same prices at equilibrium do not imply same profits, as the operators can have different costs (C_i and C_j) and thus can make different investment decisions (B_i and B_j) as shown next.

Note that our equilibrium results in medium investment regime are consistent with the well-known *Edgeworth paradox* [68] in economics. Edgeworth paradox describes a situation where two players cannot reach a state of equilibrium with pure strategies. Each operator faces capacity constraints when determining pricing decisions in Stage II. The choice of both operators charging zero prices is not an equilibrium in the medium investment regime, since at least one operator can raise its price and obtain non-zero revenue. Nor is the case where one operator charges less the other an equilibrium, since the lower price operator can profitably raise its price towards the other. Nor is the case where both operators charge the same positive price, since at least one operator can lower its price slightly and increase its profit.

The above non-equilibrium cases will not happen in the low investment regime where operators have very limited resources. This is because that in the low investment regime no operator can satisfy the whole demand alone, and thus it is possible for the two operators to share the market at the equilibrium.

Also, these non-equilibrium cases will not happen in the high investment regime, where both operators have more resources than users' total demand even the price is zero. In this regime, we can ignore the resource constraints (similar to the Bertrand competition) and the zero price equilibrium is the same as the Bertrand paradox [69].

In the Bertrand paradox, either operator deviating from zero price cannot attract any demand from its competitor who can already serve all users.

3.3.3 Operators' Leasing Strategies in Stage I

In Stage I, the operators need to decide the leasing amounts (B_i, B_j) to maximize their profits. Based on Theorem 3.1, we only need to consider the case where the total bandwidth of both the operators is no larger than Ge^{-2}. We emphasize that the analysis of Stage I is not limited to the case of low investment regime; we actually also consider the medium investment regime and the high investment regime. The key observation is that an SPE will not include any investment decisions (B_i, B_j) in the medium investment regime, as it will not lead to a pricing equilibrium in Stage II. Moreover, any investment decisions in the high investment regime lead to zero operator revenues and are strictly dominated by any decisions in low investment regime. After the above analysis, the operators only need to consider possible equilibria in the low investment regime in Stage I.

Game 2 (Investment Game) *The competition between the two operators in Stage I can be modeled as the following game:*

- *Players: two operators i and j.*
- *Strategy space: the operators will choose (B_i, B_j) from the set $\mathscr{B} = \{(B_i, B_j): B_i + B_j \leq Ge^{-2}\}$. Notice that the strategy space is coupled across the operators, but the operators do not cooperate with each other.*
- *Payoff function: the operators want to maximize their profits in (3.8) and (3.9), respectively.*

At an equilibrium of the investment game, (B_i^*, B_j^*), each operator has maximized its payoff assuming that the other operator chooses the equilibrium investment, i.e.,

$$B_i^* = \arg\max_{0 \leq B_i \leq Ge^{-2} - B_j^*} \pi_{II,i}(B_i, B_j^*), \quad i = 1, 2, i \neq j.$$

To calculate the investment equilibria of Game 2, we can first calculate operator i's best response given operator j's (not necessarily equilibrium) investment decision, i.e.,

$$B_i^*(B_j) = \arg\max_{0 \leq B_i \leq Ge^{-2} - B_j} \pi_{II,i}(B_i, B_j), \quad i = 1, 2, i \neq j.$$

By looking at operator i's profit in (3.8), we can see that a larger investment decision B_i will lead to a smaller price. The best choice of B_i will achieve the best tradeoff between a large bandwidth and a small price.

After obtaining best investment responses of duopoly, we can then calculate the investment equilibria, given different costs C_i and C_j.

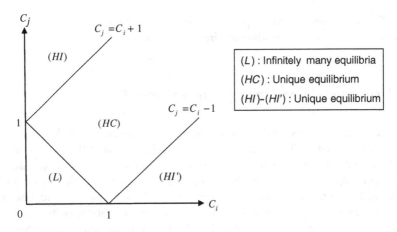

Fig. 3.4 Leasing equilibrium types in different (C_i, C_j)

Theorem 3.2 *The duopoly investment (leasing) equilibria in Stage I are summarized as follows.*

- Low Costs Regime $(0 < C_i + C_j \leq 1$, as region (L) in Fig. 3.4): *there exists infinitely many investment equilibria characterized by*

$$B_i^* = \rho G e^{-2}, \ B_j^* = (1 - \rho)G e^{-2}, \tag{3.10}$$

where ρ can be any value that satisfies

$$C_j \leq \rho \leq 1 - C_i. \tag{3.11}$$

The operators' profits are

$$\pi_{I,i}^L = B_i^*(1 - C_i),$$

$$\pi_{I,j}^L = B_j^*(1 - C_j),$$

where "L" denotes the low costs regime.

- High Comparable Costs Regime $(C_i + C_j > 1$ and $|C_j - C_i| \leq 1$, as region (HC) in Fig. 3.4): *there exists a unique investment equilibrium*

$$B_i^* = \frac{(1 + C_j - C_i)G}{2} e^{-\frac{C_i + C_j + 3}{2}}, \tag{3.12}$$

$$B_j^* = \frac{(1 + C_i - C_j)G}{2} e^{-\frac{C_i + C_j + 3}{2}}. \tag{3.13}$$

The operators' profits are

$$\pi_{I,i}^{HC} = \left(\frac{1 + C_j - C_i}{2}\right)^2 Ge^{-\left(\frac{C_i + C_j + 3}{2}\right)},$$

$$\pi_{I,j}^{HC} = \left(\frac{1 + C_i - C_j}{2}\right)^2 Ge^{-\left(\frac{C_i + C_j + 3}{2}\right)},$$

where "HC" denotes the high comparable costs regime.
- High Incomparable Costs Regime ($C_j > 1 + C_i$ or $C_i > 1 + C_j$, *as regions*
  (HI) and (HI') in Fig. 3.4): For the case of $C_j > 1 + C_i$, *there exists a unique*
  investment equilibrium with

$$B_i^* = Ge^{-(2+C_i)}, \quad B_j^* = 0,$$

i.e., operator i acts as the monopolist and operator j is out of the market. The
operators' profits are

$$\pi_{I,i}^{HI} = Ge^{-(2+C_i)}, \quad \pi_{I,j}^{HI} = 0,$$

where "HI" denotes the high incomparable costs. The case of $C_i > 1 + C_j$ *can*
be analyzed similarly.

The proof of Theorem 3.2 is given in Appendix 3.8.2. Let us further discuss the properties of the investment equilibrium in three different costs regimes.

3.3.3.1 Low Costs Regime ($0 < C_i + C_j \leq 1$)

In this case, both the operators have very low costs. It is the best response for each operator to lease as much as possible. However, since the strategy set in the Investment Game is coupled across the operators (i.e., $\mathcal{B} = \{(B_i, B_j) : B_i + B_j \leq Ge^{-2}\}$), there exist infinitely many ways for the operators to achieve the maximum total leasing amount Ge^{-2}. We can further identify the focal point, i.e., the equilibrium that the operators will agree on without prior communications [82]. The details can be found in our online technical report [90].

3.3.3.2 High Comparable Costs Regime ($C_i + C_j > 1$ and $|C_j - C_i| \leq 1$)

First, the high costs discourage the operators from leasing aggressively, thus the total investment is less than Ge^{-2}. Second, the operators' costs are comparable, and thus the operator with the slightly lower cost does not have sufficient power to drive the other operator out of the market.

3.3.3.3 High Incomparable Costs Regime $(C_j > 1 + C_i$ or $C_i > 1 + C_j)$

First, the costs are high and thus the total investment of two operators is less than Ge^{-2}. Second, the costs of the two operators are so different that the operator with the much higher cost is driven out of the market. As a result, the remaining operator thus acts as a monopolist.

3.4 Equilibrium Summary

Based on the discussions in Sect. 3.3, we summarize the equilibria of the three-stage game in Table 3.2, which includes the operators' investment decisions, pricing decisions, and the resource allocation to the users. Without loss of generality, we assume $C_i \leq C_j$ in Table 3.2. The equilibrium for $C_i > C_j$ can be decribed similarly.

Several interesting observations are as follows.

Observation 7 *The operators' equilibrium investment decisions B_i^* and B_j^* are linear in the users' aggregate wireless characteristics $G \left(= \sum_{k \in \mathcal{K}} g_k = \sum_{k \in \mathcal{K}} P_k^{max} h_k / n_0 \right)$.*

This shows that the operators' total investment increases with the user population, users' channel gains, and users' transmission powers.

Observation 8 *The symmetric equilibrium price $p_i^* = p_j^*$ does not depend on users' wireless characteristics.*

Observations 7 and 8 are closely related. Since the total investment is linearly proportional to the users' aggregate characteristics, the "average" equilibrium resource allocation per user is "constant" and does not depend on the user population. Since resource allocation is determined by the price, this means that the price is also independent of the user population and wireless characteristics.

Observation 9 *The operators can determine different equilibrium leasing and pricing decisions by observing some linear thresholds in Figs. 3.3 and 3.4.*

For equilibrium investment decisions in Stage I, the feasible set of investment costs can be divided into three regions by simple linear thresholds as in Fig. 3.4. As leasing costs increase, operators invest less aggressively; as the leasing cost difference increases, the operator with a lower cost gradually dominates the spectrum market. For the equilibrium pricing decisions, the feasible set of leasing bandwidths is also divided into three regions by simple linear thresholds as well. A meaningful pricing equilibrium exists only when the total available bandwidth from the two operators is no larger than a threshold (see Fig. 3.3).

Observation 10 *Each user k's equilibrium demand is positive, linear in its wireless characteristic g_k, and decreasing in the price. Each user k achieves the same SNR independent of g_k, and obtains a payoff linear in g_k.*

Table 3.2 Operators' and Users' Behaviors at Equilibria (assuming $C_i \leq C_j$)

Costs regimes	Low costs: $C_i + C_j \leq 1$	High comparable costs: $C_i + C_j > 1$ and $C_j - C_i \leq 1$	High incomparable costs: $C_j > 1 + C_i$
Number of equilibria	Infinite	Unique	Unique
Investment equilibria (B_i^*, B_j^*)	$(\rho Ge^{-2}, (1-\rho)Ge^{-2})$, with $C_j \leq \rho \leq (1 - C_i)$	$\left(\frac{(1+C_j-C_i)G}{2e^{\frac{C_i+C_j+3}{2}}}, \frac{(1+C_i-C_j)G}{2e^{\frac{C_i+C_j+3}{2}}}\right)$	$(Ge^{-(2+C_i)}, 0)$
Pricing equilibrium (p_i^*, p_j^*)	$(1,1)$	$\left(\frac{C_i+C_j+1}{2}, \frac{C_i+C_j+1}{2}\right)$	$(1 + C_i, N/A)$
Profits $(\pi_{I,i}, \pi_{I,j})$	$\pi_{I,i}^L = \rho(1 - C_i)Ge^{-2}$, $\pi_{I,j}^L = (1 - \rho)(1 - C_j)Ge^{-2}$	$\pi_{I,i}^{HC} = \left(\frac{1+C_j-C_i}{2}\right)^2 Ge^{-\left(\frac{C_i+C_j+3}{2}\right)}$, $\pi_{I,j}^{HC} = \left(\frac{1+C_i-C_j}{2}\right)^2 Ge^{-\left(\frac{C_i+C_j+3}{2}\right)}$	$\pi_{I,i}^{HI} = Ge^{-(2+C_i)}$, $\pi_{I,j}^{HI} = 0$
User k's bandwidth demand	$g_k e^{-2}$	$g_k e^{-\frac{C_i+C_j+3}{2}}$	$g_k e^{-(2+C_i)}$
User k's SNR	e^2	$e^{\frac{C_i+C_j+3}{2}}$	e^{2+C_i}
User k's payoff	$g_k e^{-2}$	$g_k e^{-\left(\frac{C_i+C_j+3}{2}\right)}$	$g_k e^{-(2+C_i)}$

Observation 10 shows that the users receive fair resource allocation and service quality. Such allocation does not depend on the wireless characteristics of the other users.

Observation 11 *In the High Incomparable Costs Regime, users' equilibrium SNR increases with the costs C_i and C_j, and the equilibrium payoffs decrease with the costs.*

As the costs C_i and C_j increase, the pricing equilibrium ($p_i^* = p_j^*$) increases to compensate the loss of the operators' profits due to increased costs. As a result, each user will purchase less bandwidth from the operators. Since a user spreads its total power across the entire allocated bandwidth, a smaller bandwidth means a higher SNR but a smaller payoff. Finally, we observe that the users achieve a high SNR at the equilibrium. The minimum equilibrium SNR that users achieve among the three costs regime is e^2. In this case, the ratio between the high SNR approximation and Shannon capacity, $\ln(\text{SNR})/\ln(1+\text{SNR})$, is larger than 94 %. This validates our assumption on the high SNR regime. The next section, on the other hand, shows that most of the insights remain valid in the general SNR regime.

3.4.1 How Network Dynamics Affect Equilibrium Decisions

Our analysis so far has not considered network dynamics, as we have focused on a single time slot where an operator knows users' channel conditions through proper feedback mechanisms. In this subsection, we will look at how the equilibrium results in Table 3.2 change over multiple time slots with the network dynamics. Note that operators still have the complete network information in each time slot. Users are myopic in the sense that they do not take into account the effect of time-varying network parameters on future prices when they determine bandwidth demand in the current time slot.

First, we consider the case where the spectrum available for leasing changes over time. Intuitively, when a primary operator faces a strong demand from its own primary users, it will have less spectrum resource for the virtual operator and will set a higher leasing cost. Here, we look at the case where operators' leasing costs C_i and C_j change over time according to some Markov decision processes. We write two costs as $C_i(t)$ and $C_j(t)$ to emphasize their dependancies in time. For the illustration purpose, we consider three possible values for both $C_i(t)$ and $C_j(t)$: 0.4, 0.8, and 2, and the transition probabilities (same for two operators) are shown in Fig. 3.5.

Figure 3.6 shows how costs $C_i(t)$ and $C_j(t)$, equilibrium leasing decisions B_i^* and B_j^*, and pricing decisions p_i^* and p_j^* change over time. Here we represent a price N/A in Table 3.2 as a zero price. This means that whenever we see a zero price in the figure, the corresponding operator does not participate in the game and the other operator becomes the monopoly in the market. We observe that as an operator's leasing cost increases, its leasing amount decreases. The operator with a lower cost will lease more and will become the monopolist if its cost is much lower

Fig. 3.5 Transition matrix of $C_i(t)$ and $C_j(t)$ over time slots

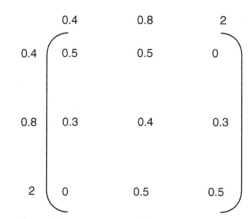

Fig. 3.6 Costs, equilibrium bandwidth and pricing decisions as functions of time slots. Here we fix ρ to be 0.5

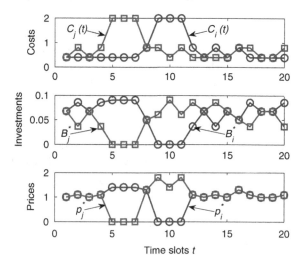

than its competitor (i.e., with $|C_j(t) - C_i(t)| > 1$ in the high incomparable costs regime). In this case, its competitor decides not to lease. As costs increase, operators' symmetric prices tend to increase to compensate the costs. When two costs are low (with $C_i(t) + C_j(t) \leq 1$), both operators announce the same high price.

Second, we can consider the dynamics of users' channel gains and their population over time. Users' channel gains may follow, for example, different Rayleigh distributions. Also, there can be users departing or entering the network over different time slots. As a result, users' aggregate wireless characteristics G will change over time. Table 3.2 has clearly shown that operators' leasing amounts and profits will change proportionally to G. But the equilibrium prices will not be affected, since operators will balance their leasing amounts with users' demands. For the sake of space, we will not show additional plots for this case.

Fig. 3.7 Pricing equilibrium
types in different (B_i, B_j)
regions for general SNR
regime

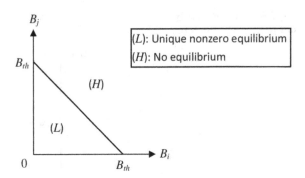

3.5 Equilibrium Analysis Under General SNR Regime

In Sects. 3.3 and 3.4, we computed the equilibria of the three-stage game based on
the high SNR assumption in (3.2), and obtained five important observations (Observations 1–5). The high SNR assumption enables us to obtain closed-form solutions
of the equilibria analysis and clear engineering insights.

In this section, we further consider the more general SNR regime where a user's
rate is computed according to (3.1) instead of (3.2). We will follow a similar backward induction analysis, and extend Observations 7, 8, 10, 11, and pricing threshold
structure of Observation 9 to the general SNR regime.

We first examine the pricing equilibrium in Stage II.

Theorem 3.3 *Define* B_{th}: $= 0.462G$. *The pricing equilibria in the general SNR
regime are as follows.*

- *Low Investment Regime* $(B_i + B_j \leq B_{th}$ *as in region* (L) *of* Fig. 3.7): *there exists
  a unique pricing equilibrium*

$$p_i^*(B_i, B_j) = p_j^*(B_i, B_j) = \ln\left(1 + \frac{G}{B_i + B_j}\right) - \frac{G}{B_i + B_j + G}. \tag{3.14}$$

The operators' profits in Stage II *are*

$$\pi_i(B_i, B_j) = B_i \left[\ln\left(1 + \frac{G}{B_i + B_j}\right) - \frac{G}{B_i + B_j + G} - C_i\right], \tag{3.15}$$

$$\pi_j(B_i, B_j) = B_j \left[\ln\left(1 + \frac{G}{B_i + B_j}\right) - \frac{G}{B_i + B_j + G} - C_j\right]. \tag{3.16}$$

- *High Investment Regime* $(B_i + B_j > B_{th}$ *as in region* (H) *of* Fig. 3.7): *there is
  no pricing equilibrium.*

Proof of Theorem 3.3 is given in Ref. [90]. This result is similar to Theorem 3.1 in the high SNR regime, and shows that the pricing equilibrium in the general SNR regime still has a *threshold structure* in Observation 9.

Unlike Theorem 3.1, here we only have two investment regimes. The "high investment regime" in Theorem 3.1 is gone, and the "medium investment regime" in Theorem 3.1 corresponds to the high investment regime here. Intuitively, the high SNR assumption in Sect. 3.3 requires each user to demand relatively small amount of bandwidth to spread its transmission power efficiently, thus the users' total demand is not elastic to prices and is always upper-bounded by Ge^{-1} in Fig. 3.3. But in the general SNR case, users' demands are elastic to prices and is no longer upper-bounded. Hence, we only have two regimes here. For more details, please refer to Ref. [90].

Based on Theorem 3.3, we are ready to prove Observations 7, 8, 10, and 11 in the general SNR regime.

Theorem 3.4 *Observations* 7, 8, 10, *and* 11 *in* Sect. 3.4 *still hold for the general SNR regime.*

Proof of Theorem 3.4 is given in Ref. [90].

3.6 Impact of Operator Competition

We are interested in understanding the impact of operator competition on the operators' profits and the users' payoffs. As a benchmark, we will consider the *coordinated* case where both operators jointly make the investment and pricing decisions to maximize their total profit. In this case, there does not exists competition between the two operators. However, it is still a Stackelberg game between a single decision maker (representing both operators) and the users. Then we will compare the equilibrium of this Stackelberg game with that of the duopoly game as in Sect. 3.4.

3.6.1 Maximum Profit in the Coordinated Case

We analyze the coordinated case following a three stage model as shown in Fig. 3.8. Compared with Fig. 3.2, the key difference here is that a single decision maker makes the decisions for two operators in both Stages I and II. In other words, the two operators coordinate with each other.

Again we use backward induction to analyze the three-stage game. The analysis of Stage III in terms of the spectrum allocation among the users is the same as in Sect. 3.3.1 (still assuming the high SNR regime), and we focus on Stages II and I. Without loss of generality, we assume that $C_i \leq C_j$.

In Stage II, the decision maker maximizes the following total profit T_π by determining prices p_i and p_j:

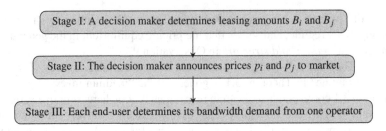

Fig. 3.8 The three-stage Stackelberg game for the coordinated operators

$$T_\pi(B_i, B_j, p_i, p_j) = \pi_i(B_i, B_j, p_i, p_j) + \pi_j(B_i, B_j, p_i, p_j),$$

where $\pi_i(B_i, B_j, p_i, p_j)$ is given in (3.4) and $\pi_j(B_i, B_j, p_i, p_j)$ can be obtained similarly.

Theorem 3.5 *In Stage II, the optimal pricing decisions for the coordinated operators are as follows:*

- *If $B_i > 0$ and $B_j = 0$, then operator i is the monopolist and announces a price*

$$p_i^{co}(B_i, 0) = \ln\left(\frac{G}{B_i}\right) - 1. \qquad (3.17)$$

Similar result can be obtained if $B_i = 0$ and $B_j > 0$.
- *If $\min(B_i, B_j) > 0$, then both operator i and j announce the same price*

$$p_i^{co}(B_i, B_j) = p_j^{co}(B_i, B_j) = \ln\left(\frac{G}{B_i + B_j}\right) - 1.$$

Proof of Theorem 3.5 can be found in Ref. [90]. Theorem 3.5 shows that both operators will act together as a monopolist in the pricing stage.

Now let us consider Stage I, where the decision maker determines the leasing amounts B_i and B_j to maximize the total profit:

$$\max_{B_i, B_j \geq 0} T_\pi(B_i, B_j) = \max_{B_i, B_j \geq 0} B_i(p_i^{co}(B_i, B_j) - C_i) + B_j(p_j^{co}(B_i, B_j) - C_j),$$
$$(3.18)$$

where $p_i^{co}(B_i, B_j)$ and $p_j^{co}(B_i, B_j)$ are given in Theorem 3.5. In this case, operator j will not lease (i.e., $B_j^{co} = 0$) as operator i can lease with a lower cost. Thus the optimization problem in (3.18) degenerates to

$$\max_{B_i \geq 0} T_\pi(B_i) = \max_{B_i \geq 0} B_i(p_i^{co}(B_i, 0) - C_i).$$

This leads to the following result.

Theorem 3.6 *In Stage I, the optimal investment decisions for the coordinated operators are*

$$B_i^{co}(C_i, C_j) = Ge^{-(2+C_i)}, \quad B_j^{co}(C_i, C_j) = 0, \qquad (3.19)$$

and the total profit is

$$T_\pi^{co}(C_i, C_j) = Ge^{-(2+C_i)}.$$

3.6.2 Impact of Competition on Operators' Profits

Let us compare the total profit obtained in the competitive duopoly case (Theorem 3.2) and the coordinated case (Theorem 3.6).

3.6.2.1 Low Costs Regime $(0 < C_i + C_j \leq 1)$

First, the total equilibrium leasing amount in the duopoly case is $B_i^* + B_j^* = Ge^{-2}$, which is larger than the total leasing amount $Ge^{-(2+C_i)}$ in the coordinated case. In other words, operator competition leads to a more aggressive overall investment. Second, the total profit at the duopoly equilibria is

$$T_\pi^L(C_i, C_j, \rho) = [\rho(1 - C_i) + (1 - \rho)(1 - C_j)]Ge^{-2},$$

where ρ can be any real value in the set of $[C_j, 1 - C_i]$. Each choice of ρ corresponds to an investment equilibrium and there are infinitely many equilibria in this case as shown in Theorem 3.2. The minimum profit ratio between the duopoly case and the coordinated case optimized over ρ is

$$\mathrm{T}_\pi \mathrm{R}^L(C_i, C_j) \triangleq \min_{\rho \in [C_j, 1-C_i]} \frac{T_\pi^L(C_i, C_j, \rho)}{T_\pi^{co}(C_i, C_j)}.$$

Since $T_\pi^L(C_i, C_j, \rho)$ is increasing in ρ, the minimum profit ratio is achieved at

$$\rho^* = C_j.$$

This means

$$\mathrm{T}_\pi \mathrm{R}^L(C_i, C_j) = [C_j(1 - C_i) + (1 - C_j)^2]e^{C_i}. \qquad (3.20)$$

Although (3.20) is a non-convex function of C_i and C_j, we can numerically compute the minimum value over all possible values of costs in this regime

$$\min_{(C_i, C_j): 0 < C_i + C_j \leq 1} \mathrm{T}_\pi \mathrm{R}^L(C_i, C_j) = \lim_{\epsilon \to 0} \mathrm{T}_\pi \mathrm{R}^L(\epsilon, 0.5 + \epsilon) = 0.75.$$

This means that the total profit achieved at the duopoly equilibrium is at least 75 % of the total profit achieved in the coordinated case under any choice of cost parameters in the Low Costs Regime.

3.6.2.2 High Comparable Costs Regime ($C_i + C_j > 1$ and $C_j - C_i \leq 1$)

First, the total duopoly equilibrium leasing amount is $B_i^* + B_j^* = Ge^{-\left(\frac{C_i+C_j+3}{2}\right)}$ which is greater than $Ge^{-(2+C_i)}$ of the coordinated case. Again, competition leads to a more aggressive overall investment. Second, the total profit of duopoly is

$$T_\pi^{HC}(C_i, C_j) = \frac{1 + (C_j - C_i)^2}{2} Ge^{-\frac{C_i+C_j+3}{2}}.$$

And the profit ratio is

$$T_\pi R^{HC}(C_i, C_j) \quad \triangleq \quad \frac{T_\pi^{HC}(C_i, C_j)}{T_\pi^{co}(C_i, C_j)} \quad = \quad \frac{1 + (C_j - C_i)^2}{2} e^{\frac{1-(C_j-C_i)}{2}},$$

which is a function of the cost difference $C_j - C_i$. Let us write it as $T_\pi R^{HC}(C_j - C_i)$. We can show that it is a convex function and achieves its minimum at

$$\min_{(C_i,C_j):C_i+C_j>1,0\leq C_j-C_i\leq 1} T_\pi R^{HC}(C_j - C_i) \quad = \quad T_\pi R^{HC}(2 - \sqrt{3}) \quad = \quad 0.773.$$

3.6.2.3 High Incomparable Costs Regime ($C_j - C_i > 1$)

In this case, only one operator leases a positive amount at the duopoly equilibrium and achieves the same profit as in the coordinated case. The profit ratio is 1.

We summarize the above results as follows.

Theorem 3.7 (*Operators' Profit Loss*). *Comparing with the coordinated case, the operator competition leads to a maximum total profit loss of* 25 % *in the low costs regime.*

Since low leasing costs lead to aggressive leasing decisions and thus intensive competitions among operators, it is not surprising to see that the maximum profit loss happens in the low cost regime. For detailed discussions on the relationship between the profit loss and the costs, see our online technical report [90].

3.6.3 Impact of Competition on Users' Payoffs

Theorem 3.8 *Comparing with the coordinated case, users obtain same or higher payoffs under the operators' competition.*

By substituting (3.19) into (3.17), we obtain the optimal price in the coordinated case as $1 + C_i$. This means that user k's payoff equals to $g_k e^{-(2+C_i)}$ in all three costs regimes. According to Table 3.2, users in the duopoly competition case have the same payoffs as in coordinated case in the high incomparable costs regime. The payoffs are larger in the other two costs regimes with the competitor competition. The intuition is that operator competition in those two regimes leads to aggressive investments, which results in lower prices and higher user payoffs.

3.7 Summary

Dynamic spectrum leasing enables the secondary network operators to quickly obtain the unused resources from the primary operaotr and provide flexible services to secondary end-users. This chapter studies the competition between two secondary operators and examines the operators' equilibrium investment and pricing decisions as well as the users' corresponding achieved service quality and payoffs.

We model the economic interactions between the operators and the users as a three-stage dynamic game. Our appropriate OFDM-based spectrum sharing model captures the wireless heterogeneity of the users in terms of maximum transmission power levels and channel gains. The two operators engage in investment and pricing competitions with asymmetric costs. We have discovered several interesting features of the game's equilibria. For example, the operators can determine different equilibrium leasing and pricing decisions by observing some linear thresholds. We also study the impact of operator competition on operators' total profit loss and the users' payoff increases. Compared with the coordinated case where the two operators cooperate to maximize their total profit, we show that at the maximum profit loss due to competition is no larger than 25 %. We also show that the users always benefit from competition by achieving the same or better payoffs. Although we have focused on the high SNR regime when obtaining closed-form solutions, we show that most engineering insights summarized in Sect. 3.4 still hold in the general SNR regime. Due to the page limit, more detailed discussions and all proofs can be found in our paper [90].

3.8 Appendix

3.8.1 Proof of Theorem 3.1

Assume, without loss of generality, that $B_i \leq B_j$. Based on Proposition 3.1, in the following analysis we examine all possible (B_i, B_j) regions labeled (a)–(f) in Fig. 3.9, and check if there exists a symmetric pricing equilibrium (i.e., $p_i^* = p_j^*$) in each region.

(a) If $B_j \geq B_i \geq Ge^{-1}$, both the operators have adequate bandwidths to cover the total preferred demand which reaches its maximum Ge^{-1} at zero price.

- if $p_i^* = p_j^* > 0$, then operator i attracts and realizes half of the total preferred demand. But when operator i slightly decreases its price, it attracts and realizes the total preferred demand, and thus doubles its revenue.
- if $p_i^* = p_j^* = 0$, any operator can not attract or realize any preferred demand by unilaterally deviating from (increasing) its price.

Hence, $p_i^* = p_j^* = 0$ is the unique equilibrium in region (a).

(b–c) If $B_i \leq Ge^{-2} < Ge^{-1} \leq B_j$ or $Ge^{-2} < B_i < Ge^{-1} \leq B_j$, operator j has adequate bandwidth while operator i only has limited bandwidth.

- if $p_i^* = p_j^* > 0$, then operator j will slightly reduce its price to attract and realize the total preferred demand.
- if $p_i^* = p_j^* = 0$, then operator j will increase its price and still have positive realized demand. This is because operator i does not have enough supply to satisfy the total preferred demand.

Hence, there doesn't exist an equilibrium in regions $(b–c)$.

(d–e) If $Ge^{-2} \leq B_i \leq B_j < Ge^{-1}$ or $B_i \leq Ge^{-2} \leq B_j < Ge^{-1}$, we have shown in the proof of Proposition 3.1 that possible pricing equilibrium will

Fig. 3.9 Different (B_i, B_j) regions

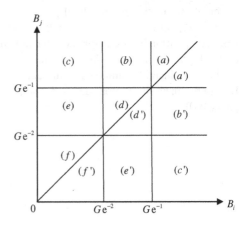

not exceed 1. We find possible pricing equilibrium given operator j's leasing amount.

- if $p_i^* = p_j^* > \ln\left(\frac{G}{B_j}\right) - 1$, then operator j has enough bandwidth to cover the total preferred demand and it will slightly decrease its price to attract a larger preferred demand.
- if $p_i^* = p_j^* \leq \ln\left(\frac{G}{B_j}\right) - 1$, then the operator j has limited bandwidth and it will make decision depending on operator i's supply.
  - if $B_i \leq Ge^{-(1+p_j^*)}/2$, then the operator j will slightly decrease its price if $B_i + B_j > Ge^{-(1+p_j^*)}$, or increase its price to 1 if $B_i + B_j \leq Ge^{-(1+p_j^*)}$.
  - if $B_i > Ge^{-(1+p_j^*)}/2$, then the operator j will slightly reduce its price.

Hence, there doesn't exist a pricing equilibrium in regions $(d{-}e)$.

(f) If $B_i \leq B_j \leq Ge^{-2}$, we will first show that total supply equals total preferred demand at any possible equilibrium (i.e., $p_i^* = p_j^* = \ln\left(\frac{G}{B_i+B_j}\right) - 1$).

- Suppose that at an equilibrium $p_i^* = p_j^* < \ln\left(\frac{G}{B_i+B_j}\right) - 1$ and thus the total supply is *less* than the total preferred demand. Then operator j will slightly increase its price without changing much its realized demand, and thus receive a greater revenue.
- Suppose that at an equilibrium $p_i^* = p_j^* \geq \ln\left(\frac{G}{B_i+B_j}\right) - 1$ and thus the total supply is *greater* than the total preferred demand. Thus we have $B_j > Ge^{-(1+p_j^*)}/2$. Operator j will slightly reduce its price to attract much more preferred demand and receive a greater revenue.

Thus we have $p_i^* = p_j^* = \ln\left(\frac{G}{B_i+B_j}\right) - 1$ at any possible equilibrium. Then we check if such (p_i^*, p_j^*) is an equilibrium for the following two cases.

- If $B_i + B_j > Ge^{-2}$, then we have $p_i^* = p_j^* < 1$. Since operator j already has its individual supply equal to its realized demand, then operator i acts as a monopolist serving its own users in the monopolist's high investment regime in the proof of Proposition 3.1. Then operator i will increase its price to 1.
- If $B_i + B_j \leq Ge^{-2}$, then we have $p_i^* = p_j^* \geq 1$. Each operator acts as a monopolist serving its own users in the monopolist's low investment regime in the proof of Proposition 3.1. And it's optimal for each operator to stick with its current price.

Thus there exists a unique pricing equilibrium $p_i^* = p_j^* = \ln\left(\frac{G}{B_i+B_j}\right) - 1$ for the low investment regime $B_i + B_j \leq Ge^{-2}$ in region (f).

The same results can be extended to symmetric regions $(a'){-}(f')$ in Fig. 3.9. ∎

Fig. 3.10 Different (C_i, C_j) regions

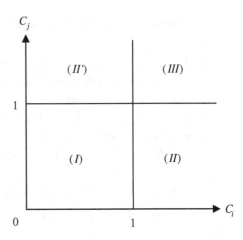

C_j

3.8.2 Proof of Theorem 3.2

The best investment response of operator i is summarized in Table 3.3 with detailed proof in Ref. [90]. An investment equilibrium (B_i^*, B_j^*) corresponds to a fixed iteration point of two functions $B_i^*(B_j)$ and $B_j^*(B_i)$. In the following analysis, we examine all possible costs (C_i, C_j) regions labeled (I)–(III) in Fig. 3.10, and check if there exists any equilibrium in each region.

(I) If $C_i \leq 1$ and $C_j \leq 1$, both the operators are in low individual cost regime.

- If $B_i^* \geq C_j Ge^{-2}$ and $B_j^* \geq C_i Ge^{-2}$, there exist infinitely many investment equilibria characterized by (3.10) and (3.11). Since $B_i^* \geq C_j Ge^{-2}$ and $B_j^* \geq C_i Ge^{-2}$, $C_i + C_j \leq 1$ is further required for existence of equilibria.
- If $B_i^* < C_j Ge^{-2}$ and $B_j^* \geq C_i Ge^{-2}$, then by solving equations $B_i^*(B_j^*) = Ge^{-2} - B_j^*$, and

Table 3.3 Best Investment Response $B_i^*(B_j)$ of Operator i in Stage I

Response $B_i^*(B_j)$	Low individual cost $0 < C_i \leq 1$	High individual cost $C_i > 1$
Small competitor's decision $B_j < C_i Ge^{-2}$	The solution to $\partial \pi_{II,i}(B_i, B_j)/\partial B_i = 0$	N/A
Large competitor's decision $B_j \geq C_i Ge^{-2}$	$Ge^{-2} - B_j$	N/A
Small competitor's decision $B_j < Ge^{-(1+C_i)}$	N/A	The solution to $\partial \pi_{II,i}(B_i, B_j)/\partial B_i = 0$
Large competitor's decision $B_j \geq Ge^{-(1+C_i)}$	N/A	0

$$\frac{\partial \pi_{II,j}(B_i, B_j)}{\partial B_j} \Big|_{B_i=B_i^*, B_j=B_j^*} = 0,$$

we have $B_i^* = C_j Ge^{-2}$ and $B_j^* = (1 - C_j)Ge^{-2}$. But the value of B_i^* is not smaller than $C_j Ge^{-2}$.

- If $B_i^* \geq C_j Ge^{-2}$ and $B_j^* < C_i Ge^{-2}$, we can also show that there does not exist any equilibrium in this case by a similar argument as above.
- If $B_i^* < C_j Ge^{-2}$ and $B_j^* < C_i Ge^{-2}$, then by solving equations

$$\frac{\partial \pi_{II,i}(B_i, B_j)}{\partial B_i} \Big|_{B_i=B_i^*, B_j=B_j^*} = 0,$$

$$\frac{\partial \pi_{II,j}(B_i, B_j)}{\partial B_j} \Big|_{B_i=B_i^*, B_j=B_j^*} = 0,$$

we have B_i^* in (3.12) and B_j^* in (3.13). And $C_i + C_j > 1$ is further required for existence of this equilibrium.

Hence, in region (I), there exist infinitely many equilibria satisfying (3.10) and (3.11) when $C_i + C_j \leq 1$, and there exists a unique equilibrium satisfying (3.12) and (3.13) when $C_i + C_j > 1$.

(II) If $C_i > 1$ and $0 < C_j \leq 1$, operator i is in high individual cost regime and operator j is in low individual cost regime.

- If $B_i^* \geq C_j Ge^{-2}$ and $B_j^* \geq Ge^{-(1+C_i)}$, then we have $B_i^* = 0$ and $B_j^* = Ge^{-2}$. But the value of B_i^* is not greater than $C_j Ge^{-2}$.
- If $B_i^* \geq C_j Ge^{-2}$ and $B_j^* < Ge^{-(1+C_i)}$, then by solving equations $B_j^*(B_i^*) = Ge^{-2} - B_i^*$, and

$$\frac{\partial \pi_{II,i}(B_i, B_j)}{\partial B_i} \Big|_{B_i=B_i^*, B_j=B_j^*} = 0,$$

we have $B_i^* = (1 - C_i)Ge^{-2}$ and $B_j^* = C_i Ge^{-2}$. But the value of B_j^* is not less than $Ge^{-(1+C_i)}$.

- If $B_i^* < C_j Ge^{-2}$ and $B_j^* \geq Ge^{-(1+C_i)}$, then by solving equations $B_i^*(B_j^*) = 0$, and

$$\frac{\partial \pi_{II,j}(B_i, B_j)}{\partial B_j} \Big|_{B_i=B_i^*, B_j=B_j^*} = 0,$$

we have $B_i^* = 0$ and $B_j^* = Ge^{-(2+C_j)}$. And $C_i > 1 + C_j$ is further required for existence of this equilibrium.

- $B_i^* < C_j Ge^{-2}$ and $B_j^* < Ge^{-(1+C_i)}$, then by solving equations

$$\frac{\partial \pi_{II,i}(B_i, B_j)}{\partial B_i} \Big|_{B_i=B_i^*, B_j=B_j^*} = 0,$$

$$\frac{\partial \pi_{II,j}(B_i, B_j)}{\partial B_j}\Big|_{B_i=B_i^*, B_j=B_j^*} = 0,$$

we have B_i^* in (3.12) and B_j^* in (3.13). And $C_i \le 1 + C_j$ is further required for existence of this equilibrium.

Hence, in region (II), there exists a unique investment equilibrium (B_i^*, B_j^*) satisfying (3.12) and (3.13) when $C_i \le 1 + C_j$, and there exists a unique equilibrium satisfying $B_i^* = 0$ and $B_j^* = Ge^{-(2+C_j)}$ when $C_i > 1 + C_j$.

(III) If $C_i > 1$ and $C_j > 1$, then both the operators are in high individual cost regime.

- If $B_i^* < Ge^{-(1+C_j)}$ and $B_j^* < Ge^{-(1+C_i)}$, then by solving equations

$$\frac{\partial \pi_{II,i}(B_i, B_j)}{\partial B_i}\Big|_{B_i=B_i^*, B_j=B_j^*} = 0,$$

$$\frac{\partial \pi_{II,j}(B_i, B_j)}{\partial B_j}\Big|_{B_i=B_i^*, B_j=B_j^*} = 0,$$

  we have B_i^* in (3.12) and B_j^* in (3.13). And $C_i - 1 < C_j < C_i + 1$ is further required for existence of this equilibrium.
- If $B_i^* < Ge^{-(1+C_j)}$ and $B_j^* \ge Ge^{-(1+C_i)}$, then by solving equations $B_i^*(B_j^*) = 0$, and

$$\frac{\partial \pi_j(B_i, B_j)}{\partial B_j}\Big|_{B_i=B_i^*, B_j=B_j^*} = 0,$$

  we have $B_i^* = 0$ and $B_j^* = Ge^{-(2+C_j)}$. And $C_j \le C_i - 1$ is further required for existence of this equilibrium.
- If $B_i^* \ge Ge^{-(1+C_j)}$ and $B_j^* < Ge^{-(1+C_i)}$, then we can similarly show that there exists a unique equilibrium $B_i^* = Ge^{-(2+C_i)}$ and $B_j^* = 0$ only when $C_j \ge C_i + 1$.
- If $B_i^* \ge Ge^{-(1+C_j)}$ and $B_j^* \ge Ge^{-(1+C_i)}$, then we have $B_i^* = 0$ and $B_j^* = 0$. However, the value of B_i^* is not greater than $Ge^{-(1+C_j)}$.

Hence, in region (III), there exists a unique equilibrium satisfying (3.12) and (3.13) when $C_i - 1 < C_j < C_i + 1$; there exists a unique equilibrium satisfying $B_i^* = 0$ and $B_j^* = Ge^{-(2+C_j)}$ when $C_j \le C_i - 1$; and there exists a unique equilibrium with $B_i^* = Ge^{-(2+C_i)}$ and $B_j^* = 0$ when $C_j \ge C_i + 1$. The same results can be extended to symmetric region (II') in Fig. 3.10. ∎

Chapter 4
Conclusion

This book provides an overview of cognitive mobile virtual network operators'
decisions under investment flexibility, supply uncertainty, and market competition in
cognitive radio networks. This is a new research area at the nexus of cognitive radio
engineering and microeconomics. We propose two flexible approaches, i.e., spectrum sensing and dynamic spectrum leasing, to resolve the inflexible supply problem
for the virtual operators who are traditionally stuck in long-term leasing contracts
with spectrum owners. Our focus is the virtual operator's joint spectrum investment
and service pricing decisions at a short time scale. Despite the investment flexibility, spectrum sensing would introduce supply uncertainty due to primary licensed
users' stochastic traffic. Thus we examine how to tradeoff the two flexible investment approaches under supply uncertainty. Our results show that spectrum sensing
can significantly increase a virtual operator's expected profit (up to 25 %), although
the realized profit in one time slot may decrease due to the availability of spectrum
sensing. When there are more than one operator, we analyze the operators's competition in both spectrum acquisition and service pricing to attract local users. We show
that secondary users would benefit from the operator competition, and the market
regulator should advocate such competition (e.g., by lowing the entry barrier for the
wireless market).

L. Duan et al., *Cognitive Virtual Network Operator Games*,
SpringerBriefs in Computer Science, DOI: 10.1007/978-1-4614-8890-3_4,
© The Author(s) 2013

References

1. FCC Spectrum Policy Task Force, Report of the spectrum efficiency working group. http://transition.fcc.gov/sptf/reports.html November 2002
2. FCC, ET Docket No 03–222 Notice of proposed rule making and order. December 2003
3. M.A. McHenry, P.A. Tenhula, D. McCloskey, D.A. Roberson, C.S. Hood, Chicago spectrum occupancy measurements and analysis and a long-term studies proposal, in *Proceedings of ACM the First International Workshop on Technology and Policy for Accessing, Spectrum*, (2006)
4. I.F. Akyildiz, W.-Y. Lee, M.C. Vuran, S. Mohanty, NeXt generation/dynamic spectrum access/cognitive radio wireles networks: a survey. Comput. Netw. **50**(13), 2127–2159 (2006)
5. S. Haykin, Cognitive radio: brain-empowered wireless communications. IEEE J. Sel. Areas Commun. **23**(2), 201–220 (2005)
6. Visiongain, Consumer MVNOs report 2011–2016: Designing successful strategies for monetising smartphone and LTE growth. Visiongain Report, (2011)
7. C. Raman, R. Yates, N. Mandayam, Scheduling variable rate links via a spectrum server, in *Proceedings of IEEE Symposium on New Fronteiers in Dynamic Spectrum Access Networks (DySPAN)*, (2005)
8. O. Ileri, D. Samardzija, N. Mandayam, Demand responsive pricing and competitive spectrum allocation via a spectrum server, in *Proceedings of IEE International Symposium on Information Theory*, (2005)
9. J. Kanervisto, MVNO pricing strctures in Finland, Finnish Ministry of Transport and Communications, (2005)
10. T.K. Forde, I. Macaluso, L. E. Doyle, Exclusive sharing and virtualization of the cellular network, in *Proceedings of IEEE Symposium on New Fronteiers in Dynamic Spectrum Access Networks (DySPAN)*, (2011)
11. M.B.H. Weiss, M. Altamimi, L. Cui, Spatio-temporal spectrum modeling: taxonomy and economic evaluation of context acquisition. Telecommun. Policy **36**(4), 335–348 (2012)
12. S. Chung, S. Kim, J. Lee, J. Cioffic, A game-theoretic approach to power allocation in frequency-slective Gaussian interference channels, in *Proceedings of IEEE International Symposium on Information Theory*, (2003)
13. R. Etkin, A. Parekh, D. Tse, Spectrum sharing for unlicensed bands, in *Proceedings of IEEE Symposium on New Frontiers in Dynamic Spectum Access Networks (DySPAN)*, (2005)
14. J. Huang, R. Berry, M. Honig, Spectrum sharing with distributed interference compensation, in *Proceedings of IEEE Symposium on New Frontiers in Dynamic Spectrum Access Networks (DySPAN)*, (2005)

L. Duan et al., *Cognitive Virtual Network Operator Games*,
SpringerBriefs in Computer Science, DOI: 10.1007/978-1-4614-8890-3,
© The Author(s) 2013

15. Q. Zhao, B. Sadler, A survey of dynamic spectrum access: signal processing, networking, and regulatory policy. IEEE Signal Process. Mag. **24**(3), 78–89 (May 2007)

16. Q. Zhao, Spectrum Opportunity and interference constraint in opportunistic spectrum access, in *Proceedings of IEEE International Conference on Acoustics, Speech and Signal Processing (ICASSP)*, (2007)

17. J. Mitola, Cognitive radio for flexible mobile multimedia communications, in *Proceedings of IEEE International Workshop on Mobile Multimedia Communications (MoMuC)*, (1999)

18. L.B. Le, E. Hossain, Resource allocation for spectrum underlay in cognitive radio networks. IEEE Trans. Wireless Commun. **7**(12), 5306–5315 (2008)

19. R. Menon, R.M. Buehrer, J.H. Reed, Outage probability based comparison of underlay and overlay spectrum sharing techniques, in *Proceedings of IEEE International Symposium on New Frontiers in Dynamic Spectrum Access Networks (DySPAN)*, (2005)

20. D. Hatfield, P. Weiser, Property rights in spectrun: taking the next step, in *Proceedings of IEEE Symposium on New Frontiers in Dynamic Spectrum Access Networks (DySPAN)*, (2005)

21. L. Xu, R. Tonjes, T. Paila, W. Hansmann, M. Frank, M. Albrecht, DRiVE-ing to the Internet: dynamic radio for IP services in vhicular environments, in *Proc. of The 25th Annual IEEE Conference on Local, Computer Networks*, (2000)

22. S.K. Jayaweera, T. Li, Dynamic spectrum leasing in cognitive radio networks via primary-secondary user power control games. IEEE Trans. Wireless Commun. **8**(6), 3300–3310 (Jun. 2009)

23. M.M. Buddhikot, Understanding dynamic spectrum access: models, taxonomy and challenges, in *Proceedings of IEEE Symposium on New Fronteiers in Dynamic Spectrum Access Networks (DySPAN)*, April 2007

24. G. Faulhaber, D. Farber, Spectrum management: property rights, markets and the commons, in *Proceedings of Telecommunications Policy Research Conference*, October 2003

25. F. Hou, J. Huang, Dynamic channel selection in cognitive radio network with channel heterogeneity, in *Proceedings of IEEE Global Communications Conference (Best Paper Award)*, December 2010

26. F. Wang, J. Zhu, J. Huang, Y. Zhao, Admission control and channel allocation for supporting real-time applications in cognitive radio networks, in *Proceedings of IEEE Global Communications Conference (GLOBECOM)*, December 2010

27. L.M. Law, J. Huang, M. Liu, and S.-Y.R. Li, Price of anarchy for cognitive MAC games, in *Proceedings of IEEE Global Communications Conference (GLOBECOM)*, December 2010

28. M. Wellens, A. de Baynast, P. Mahonen, Exploiting historical spectrum occupancy information for adaptive spectrum sensing, in *Proceedings of IEEE Wireless Communications and Networking Conference (WCNC)*, April 2008

29. S.-Y. Tu, K.-C. Chen, R. Prasad, Spectrum sensing of OFDMA systems for cognititve radios. in *Proceedings of The 18th Annual IEEE International Symposium on Personal, Indoor and Mobile Radio Communications (PIMRC)*, September 2007

30. G. Ganesan, Y. Li, Cooperative spectrum sensing in cognitive radio, part II: multiuser networks. IEEE Trans. Wireless Commun. **6**(6), 2204–2213 (Jun. 2007)

31. T. Yucek, H. Arslan, A survey of spectrum sensing algorithms for cognititve radio applications. IEEE Commun. Surv. Tutorials **11**(1), 116–130 (2009)

32. S. Huang, X. Liu, Z. Ding, Optimal sensing-transmission structure for dynamic spectrum access, in *Proceedings of The IEEE International Conference on Computer Communications (INFOCOM)*, (2009)

33. S. Huang, X. Liu, Z. Ding, Opportunistic spectrum access in cognitive radio networks, in *Proceedings of The IEEE International Conference on Computer Communications (INFOCOM)*, (2008)

34. O. Simeone, I. Stanojev, S. Savazzi, Y. Bar-Ness, U. Spagnolini, R. Pickholtz, Spectrum leasing to cooperating seconday ad hoc networks. IEEE J. Sel. Areas Commun. **26**(1), 203–213 (Jan. 2008)

35. J.M. Chapin, W.H. Lehr, Time-limited leases in radio systems. IEEE Commun. Mag. **45**(6), 76–82 (Jun. 2007)

36. J.M. Chapin, W.H. Lehr, The path to market success for dynamic spectrum access technology. IEEE Commun. Mag. **45**(5), 96–103 (May 2007)
37. J. Jia, Q. Zhang, Competitions and dynamics of duopoly wireless service providers in dynamic spectrum market, in *Proceedings of The ACM International Symposium on Mobile Ad Hoc Networking and Computing (MobiHoc)*, (2008)
38. M. Lev-Ram, As mobile ESPN falls, a helio rises. CNN Business 2.0 Magazine. http://money.cnn.com/2006/10/17/magazines/business2/espnmobile_mvno.biz2/index.htm (2006)
39. R. Dewenter, J. Haucap, Incentives to lease mobile virtual network operators (MVNOs). Access Pricing: Theory and Pracgtice (2006), pp. 305–325
40. C. Stevenson, G. Chouinard, Z. Lei, W. Hu, S. Shellhammer, W. Caldwell, IEEE 802.22: The first cognitive radio wireless regional area network standard. IEEE Commun. Mag. **47**(1), 130–138 (2009)
41. S. Sengupta, M. Chatterjee, An economic framework for dynamic spectrum access and service pricing. IEEE/ACM Trans. Netw. **17**(4), 1200–1213 (Aug. 2009)
42. L. Duan, A.W. Min, J. Huang, K.G. Shin, Attack prevention for collaborative spectrum sensing in cognitive radio networks. IEEE J. Sel. Areas Commun. (JSAC) **30**(9), 1658–1665 (Oct. 2012)
43. J. Jia, Q. Zhang, Q. Zhang, M. Liu, Revenue generation for truthful spectrum auction in dynamic spectrum access, in *Proceedings of The ACM International Symposium on Mobile Ad Hoc Networking and Computing (MobiHoc)*, (2009)
44. J. Huang, R.A. Berry, M.L. Honig, Auction-based spectrum sharing. Springer Mob. Netw. Appl. **11**, 405–418 (2006)
45. J. Huang, R.A. Berry, M.L. Honig, Distributed interference compensation for wireless networks. IEEE J. Sel. Areas Commun. **24**, 1074–1084 (2006)
46. M. Manshaei, M. Felegyhazi, J. Freudiger, J. Hubaux, P. Marbach, *Spectrum sharing games of network operators and cognitive radios* Cognitive Wireless Networks: Concepts, Methodologies and Visions (2007)
47. D. Niyato, E. Hossain, Competitive pricing for spectrum sharing in cognitive radio networks: dynamic game, inefficiency of Nash equilibrium, and collusion. IEEE J. Sel. Areas Commun. **26**(1), 192–202 (2008)
48. H. Inaltekin, T. Wexler, S.B. Wicker, A duopoly pricing game for wireless IP services, in *Proceedings of IEEE Communications Society Conference on Sensor, Mesh and Ad Hoc Communications and Networks (SECON)*, June 2007, pp. 600–609
49. O. Ileri, D. Samardzija, T. Sizer, N. B. Mandayam, Demand responsive pricing and competitve spectrum allocation via a spectrum server, in *Proceedings of IEEE Symposium on New Frontiers in Dynamic Spectrum Access Networks (DySPAN)*, (2005)
50. Y. Xing, R. Chandramouli, C. Cordeiro, Price dynamics in competitive agile spectrum access markets. IEEE J. Sel. Areas Commun. **25**(3), 613–621 (2007)
51. J. Zhang, Q. Zhang, Stackelberg game for utility-based cooperative cognitiveradio networks, in *Proceedings of The ACM international symposium on Mobile ad hoc networking and computing (MobiHoc)*, (2009)
52. D. Niyato, E. Hossain, Hierarchical spectrum sharing in cognitive radio: a microeconomic approach. in *Proceedings of IEEE Wireless Communications and Networking Conference (WCNC)*, 2007, pp. 3822–3826
53. D. Niyato, E. Hossain, Z. Han, Dynamics of multiple-seller and multiple-buyer spectrum trading in cognitive radio networks: a game-theoretic modeling approach. IEEE Trans. Mobile Comput. **8**(8), 1009–1022 (2009)
54. J. Jia, Q. Zhang, Bandwidth and price competitions of wireless service providers in two-stage spectrum market, in *Proceedings of IEEE International Conference on Communications (ICC)*, May 2008, pp. 4953–4957
55. V. Babich, P.H. Ritchken, A. Burnetas, Competition and diversification effects in supply chains with supplier default risk. Manuf. Serv. Operators Manange. **9**(2), 123–146 (2007)
56. S. Deo, C.J. Corbett, Cournot competition under yield uncertainty: the case of the U.S. influenza vaccine market. Manufacturing and Service Operations Management, Articles in Advance. (2008), pp. 1–14

57. B. Shou, J. Huang, Z. Li, Managing supply uncertainty under chain-to-chain competition. working paper, (2009) http://personal.ie.cuhk.edu.hk/ jwhuang/publication/chain-to-chain.pdf

58. Y.-C. Liang, Y.H. Zeng, E. Peh, A.T. Hoang, Sensing-throughput tradeoff for cognitive radio networks. IEEE Trans. Wireless Commun. **7**(4), 1326–1337 (2008)

59. Y. Pei, A. Hoang, Y.-C. Liang, Sensing-throughput tradeoff in cognitive radio networks: how frequently should spectrum sensing be carried out?, in *Proceedings of IEEE International Symposium on Personal, Indoor and Mobile Radio Communications (PIMRC)*, (2007)

60. M. Weiss, S. Delaere, W. Lehr, Sensing as a service: an exploration into practical implementations of DSA, in *Proceedings of IEEE Symposium on New Frontiers in Dynamic Spectrum Access Networks (DySPAN)*, (2010)

61. SENDORA Deliverable 2.1, Scenario descriptions and system requirements. European Union, Project number ICT-2007-216076, (2008)

62. D. Willkomm, S. Machiraju, J. Bolot, A. Wolisz, Primary user behavior in cellular networks and implications for dynamic spectrum access. IEEE Commun. Mag. **47**(3), 88–95 (2009)

63. S. Geirhofer, L. Tong, B. Sadler, Dynamic spectrum access in WLAN channels: empirical model and its stochastic analysis, in *Proceedings of The First International Workshop on Technology and Policy for Accessing, Spectrum*, (2006)

64. A. Azzalini, A note on the estimation of a distribution function and quantiles by a kernel method. Biometrika **68**(1), 326–328 (Apr. 1981)

65. H. Kim, K.G. Shin, Efficient discovery of spectrum opportunities with MAC-layer sensing in cognitive radio networks. IEEE Trans. Mobile Comput. 533–545, (2007)

66. T. Shu, M. Krunz, Throughput-efficient sequential channel sensing and probing in cognitive radio networks under sensing errors, in *Proceedings of The ACM Annual International Conference on Mobile Computing and Networking*, September 2009

67. Y. Chen, Q. Zhao, A. Swami, Distributed cognititve mac for energy-constrained opportunistic spectrum access, in *Proceedings of IEEE Military Communication Conference (MILCOM)*, (2006)

68. E. Rasmusen, *Games and information: an introduction to game theory*. (Wiley-Blackwell, Malden, 2007)

69. A. Mas-Colell, M.D. Whinston, J.R. Green, *Microeconomic Theory*. (Oxford University Press, USA, 1995)

70. E. Alpaydin, Introduction to machine learning. (The MIT Press, Cambridge, 2004)

71. L. Duan, J. Huang, B. Shou, Competition with dynamic spectrum leasing, in *Proceedings of IEEE Symposium on New Frontiers in Dynamic Spectrum Access Networks (DySPAN)*, (2010)

72. Research and Markets, The MVNO directory 2009. (Blycroft Ltd, 2009)

73. J. Huang, R.A. Berry, M.L. Honig, Auction-based spectrum sharing. ACM/Springer Mobile Netw. Appl. J. (MONET) **24**(5), 1074–1084 (2006)

74. S. Wang, P. Xu, X. Xu, S.-J. Tang, X.-Y. Li, X. Liu, ODA: Truthful online double auction for spectrum allocation in wireless networks, in *Proceedings of IEEE Symposium on New Frontiers in Dynamic Spectrum Access Networks (DySPAN)*, (2010)

75. X. Wang, Z. Li, P. Xu, Y. Xu, X. Gao, H. Chen, Spectrum sharing in cognitive radio networks C an auction-based approach. IEEE Trans. Syst. Man Cybern.C Part B Cybern. to appear

76. V. Gajic, J. Huang, B. Rimoldi, Competition of wireless providers for atomic users: Equilibrium and social optimality, in *Proceedings of Allerton Conference*, Monticello, September 2009

77. P. Maille, B. Tuffin, Analysis of price competition in a slotted resource allocation game, in *Proceedings of The IEEE International Conference on Computer Communications (INFOCOM)*, (2008)

78. G.S. Kasbekar, S. Sarkar, Spectrum pricing games with bandwidth uncertainty and spatial reuse in cognitive radio networks, in *Proceedings of The ACM International Symposium on Mobile Ad Hoc Networking and Computing (MobiHoc)*, September 2010

79. D. Niyato, E. Hossain, Z. Han, Dynamic spectrum access in ieee 802.22-based cognititve wireless networks: a game theoretic model for competitive spectrum bidding and pricing. IEEE Wireless Commun. **16**(2), 16–23 (Apr. 2009)

80. L. Duan, J. Huang, B. Shou, Cognitive mobile virtual network operator: Investment and pricing with supply uncertainty, in *Proceedings of The IEEE International Conference on Computer Communications (INFOCOM)*, (2010)

81. T.A. Weiss, F.K. Jondral, Spectrum pooling: an innovative strategy for the enhancement of spectrum efficiency. IEEE Commun. Mag. **42**, S8–S14 (Mar. 2004)

82. D. Fudenberg, J. Tirole, Game theory. (MIT Press, Cambridge, 1991)

83. H.A. Mahmoud, H. Arslan, Sidelobe suppression in ofdmbased spectrum sharing systems using adaptive symbol transition. IEEE Commun. Lett. **12**(2), 133–135 (2008)

84. J. Huang, V. Subramanian, R. Agrawal, R. Berry, Downlink scheduling and resource allocation for OFDM systems. IEEE Trans. Wireless Commun. **8**(1), 288–296 (2009)

85. J. Huang, V. Subramanian, R. Agrawal, R. Berry, Joint scheduling and resource allocation in uplink OFDM systems for broadband wireless access networks. IEEE J. Sel. Areas Commun. **27**(2), 226–234 (Feb. 2009)

86. J. Bae, E. Beigman, R.A. Berry, M.L. Honig, R. Vohra, Sequential bandwidth and power auctions for distributed spectrum sharing. IEEE J. Sel. Areas Commun. **26**(7), 1193–1203 (2008)

87. IEEE 802.16 Working Group on Broadband Wireless Access Standards, IEEE 802.16e-2005 and IEEE Std 802.16-2004/Cor1- 2005. http://www.ieee802.org/16/ (2005)

88. F. Kelly, Charging and rate control for elastic traffic. Eur. Trans. Telecommun. **8**(1), 33–37 (1997)

89. S. Shakkottai, R. Srikant, Economics of network pricing with multiple ISPs. IEEE/ACM Trans. Netw. **14**(6), 1233–1245 (2006)

90. L. Duan, J. Huang, B. Shou, Duopoly competition in dynamic spectrum leasing and pricing. IEEE Trans. Mobile Comput. **11**(11), 1706–1719 (2012). http://arxiv.org/abs/1003.5517

91. Q. Zhao, L. Tong, A. Swami, Y. Chen, Decentralized cognitive MAC for opportunistic spectrum access in ad hoc networks: a POMDP framework. IEEE J. Sel. Areas Commun. **25**(3), 589–600 (2007)

92. S. Li, J. Huang, S.-Y. R. Li, Revenue maximization for communication networks with usage-based pricing, in *Proceedings of IEEE Global Communications Conference*, December 2009

93. L. Duan, J. Huang, B. Shou, Investment and pricing with spectrum uncertainty: a cognitive operators perspective. IEEE Trans. Mobile Comput. **10**(11), (2011). http://arxiv.org/abs/0912.3089

94. C. Courcoubetis, R. Weber, Pricing communication networks. Wiley Online, Library, 2003, vol. 2

95. P. Dasgupta, E. Maskin, The existence of equilibrium in discontinuous economic games, part I: theory. Review Econ. Stud. **53**(1), 1–26 (1986)

96. R. Gibbens, R. Mason, R. Steinberg, Internet service classes under competition. IEEE J. Sel. Areas Commun. **18**(12), 2490–2498 (2000)

97. D. Abreu, On the theory of infinitely repeated games with discounting. Econometrica: J. Econometric Soc. 383–396 (1988)

98. A. Ghasemi, E.S. Sousa, Opportunistic spectrum access in fading channels through collaborative sensing. J. Commun. **2**(2), 71–82 (2007)

99. E. Peh, Y.-C. Liang, Optimization for cooperative sensing in cognitive radio networks, in *Proceedings of IEEE Wireless Communications and Networking Conference (WCNC)*, (2007)

100. W. Wang, H. Li, Y. Sun, Z. Han, CatchIt: detect malicious nodes in collaborative spectrum sensing, in *Proceedings of IEEE Global Communications Conference (GLOBECOM)*, (2009)

101. A. Azzalini, A note on the estimation of a distribution function and quantiles by a kernel method. Biometrika **68**(1), 326–328 (1981)

102. Y.-C. Liang, Y. Zeng, E. Peh, A. T. Hoang, Sensing-throughput tradeoff for cognitive radio networks. IEEE Trans. Wireless Commun. **7**(4), 1326–1337 (2008)

103. A. W. Min, K. G. Shin, On Sensing-Access Tradeoff in Cognitive Radio Networks, in *Proceedings of IEEE Symposium on New Frontiers in Dynamic Spectrum Access Networks (DySPAN)*, (2010)

104. C. Cordeiro, K. Challapali, D. Birru, IEEE 802.22: An introduction to the first wireless standard based on cognitive radios. J. Commun. **1**(1), 38–47 (2006)
105. S. Huang, X. Liu, Z. Ding, Optimal sensing-transmission structure for dynamic spectrum access, in *Proceedings of The IEEE International Conference on Computer Communications (INFOCOM)*, (2009)
106. A. Chiang, *Fundamental methods of mathematical economics* (McGraw-Hill Book Co., New York, 1967)
107. J. H. Schiller, Mobile communications. (Addison Wesley, Lebanon, 2003)
108. R.B. Myerson, *Game theory: analysis of conflict* (Harvard University Press, Cambridge, 2002)
109. J. Altmann, K. Chu, How to charge for network services- flat-rate or usage-based? Comput. Netw. **36**(5–6), 519–531 (2001)
110. L. Duan, L. Gao, J. Huang, Contract-based cooperative spectrum sharing, in *Proceedings of IEEE Symposium on New Frontiers in Dynamic Spectrum Access Networks (DySPAN)*, (2011)
111. L. Duan, L. Gao, J. Huang, Cooperative spectrum sharing: a contract-based approach. IEEE Trans. Mobile Comput. (TMC), forthcoming
112. L. Duan, *Some Economics of Cellular and Cognitive Radio Networks* (The Chinese University of Hong Kong, Hong Kong, 2012). Ph.D. Dissertation